# VIBRATIONAL SPECTROSCOPY AT HIGH EXTERNAL PRESSURES
## The Diamond Anvil Cell

# VIBRATIONAL SPECTROSCOPY AT HIGH EXTERNAL PRESSURES

## The Diamond Anvil Cell

*John R. Ferraro*
Chemistry Department
Loyola University
Chicago, Illinois

 1984

ACADEMIC PRESS, INC.
*(Harcourt Brace Jovanovich, Publishers)*
Orlando  San Diego  New York  London
Toronto  Montreal  Sydney  Tokyo

COPYRIGHT © 1984, BY ACADEMIC PRESS, INC.
ALL RIGHTS RESERVED.
NO PART OF THIS PUBLICATION MAY BE REPRODUCED OR
TRANSMITTED IN ANY FORM OR BY ANY MEANS, ELECTRONIC
OR MECHANICAL, INCLUDING PHOTOCOPY, RECORDING, OR ANY
INFORMATION STORAGE AND RETRIEVAL SYSTEM, WITHOUT
PERMISSION IN WRITING FROM THE PUBLISHER.


ACADEMIC PRESS, INC.
Orlando, Florida 32887


*United Kingdom Edition published by*
ACADEMIC PRESS, INC. (LONDON) LTD.
24/28 Oval Road, London NW1  7DX



**Library of Congress Cataloging in Publication Data**

Ferraro, John R., Date
  Vibrational spectroscopy at high external pressures.

  Includes bibliographies and index.
  1. Vibrational spectra.  I. Title.
QC454.V5F47   1984     543'.0858     83-22355
ISBN 0−12−254160−X  (alk. paper)




PRINTED IN THE UNITED STATES OF AMERICA

84 85 86 87    9 8 7 6 5 4 3 2 1

This book is dedicated to the coinventors of the diamond anvil cell—Alvin Van Valkenburg and Charles Weir.

# CONTENTS

*Preface* xi
*Acknowledgments* xiii

## CHAPTER 1  Introduction

1. Effects of Pressure 2
2. Pressure Units 5
   References 7
   Bibliography 7

## CHAPTER 2  Instrumentation

1. Introduction 9
2. High-Pressure Optical Cell 9
3. Optical Windows for Use at High Pressures 16
4. Optical Instrumentation in the IR and for Raman Experiments 17
5. Optical Link of Pressure Cell to Optical Cell 26
6. Generation of Temperatures Simultaneously with Pressure in the DAC 28
7. Complementary Measurements in the DAC 28
   References 37
   Bibliography 40

## CHAPTER 3  Pressure Calibration

1. Introduction 45
2. Methods of Pressure Calibration 48

|   |   |
|---|---|
| References | 60 |
| Bibliography | 61 |

## CHAPTER 4  Inorganic Compounds

| | |
|---|---|
| 1. Introduction | 63 |
| 2. Elements (Nonmetals) | 65 |
| 3. Type AX (II–VI and III–IV Compounds) | 67 |
| 4. Alkali Metal Halides (AB) | 73 |
| 5. Alkaline-Earth Fluorides | 81 |
| 6. Thallium Iodide | 83 |
| 7. Bihalide Salts | 84 |
| 8. $AB_2$ Halides | 85 |
| 9. Miscellaneous Systems | 90 |
| 10. Miscellaneous Inorganic Compounds | 101 |
| 11. Ionic Conductors | 107 |
| References | 115 |
| Bibliography | 118 |

## CHAPTER 5  Coordination Compounds

| | |
|---|---|
| 1. Solid-State Structural Conversions | 120 |
| References | 160 |
| Bibliography | 162 |

## CHAPTER 6  Organic and Biological Compounds

| | |
|---|---|
| 1. Organic Molecules | 164 |
| 2. Biological Compounds | 189 |
| References | 192 |
| Review Articles | 193 |
| Bibliography | 193 |

## CHAPTER 7  Special Applications

| | |
|---|---|
| 1. Geochemical and Geophysical Applications | 194 |
| 2. Electrical Conductor Applications | 208 |
| 3. Lubricant Studies | 219 |
| 4. Forensic Science | 220 |
| References | 231 |
| Bibliography | 235 |

## CHAPTER 8  Miscellaneous Applications

| | |
|---|---|
| 1. Metallic Hydrogen | 236 |
| 2. Metallic Xenon | 241 |

*Contents*

3. Cuprous Chloride—Superconductor or Not? 242
   Addendum 244
   References 244

## Additional Bibliography 247

*Index* 261

# PREFACE

In 1959, when the diamond anvil cell (DAC) was discovered by workers at the National Bureau of Standards, it became possible to conveniently study materials under pressure by using infrared spectroscopy in the midinfrared region. This technique was later extended into the far infrared, and Raman scattering experiments became possible with the advent of the laser. The compact size of the DAC made it possible to fit the cell into sample compartments of various infrared spectrophotometers, which made such experiments possible for the first time.

In 24 years of spectroscopic research with the DAC, extensive new and interesting data have been collected. The cell has been used primarily with solids, but its use with liquids has also been examined. Diverse compounds have been studied, including inorganic, organic, coordination, biological, and polymeric compounds. Many other disciplines have also benefited. Applications involving geochemistry, geophysics, forensic science, conductors, lubricants, polymers, and other areas have surfaced. The utility of the DAC in high-pressure work has been proven beyond any reasonable doubt.

It is the purpose of this book to present the effects of high pressure on the vibrational properties of materials as accomplished in a DAC. The subject matter involving this cell has expanded to such an extent that although piston–cylinder cells can perform similar experiments, at least in the near infrared, we have chosen to restrict ourselves to the DAC in order to keep the scope of this book reasonable. Furthermore, no extensive synopsis of the DAC and its achievements has previously been published.

# ACKNOWLEDGMENTS

We wish to acknowledge the contributions made by many students at the Argonne National Laboratory and Loyola University who made possible certain portions of the work presented here. The interaction and contribution of various colleagues and visiting scientists at the Argonne National Laboratory have proven particularly helpful as well.

It is with pleasure that we acknowledge the former U.S. Atomic Energy Commission, the Energy Research and Development Administration, and the present Department of Energy for supporting much of the work presented in this book. Last but not least, we wish to thank the Searle Foundation for its present support of our work.

We also wish to thank Dr. Alvin Van Valkenburg of High Pressure Diamond Optics, Inc., and Dr. Louis J. Basile of the Argonne National Laboratory for their critical reading and Mrs. Sheila Honda for her typing of the manuscript. Thanks are also in order to Dr. Van Valkenburg for the use of his computer printout of references relative to high-pressure research; Dr. Peter M. Bell of the Geophysics Laboratory; Dr. W. A. Bassett of Cornell University; Dr. W. F. Sherman of Kings College; Dr. Charles F. Midkeff of the National Laboratory Center, Bureau of Alcohol, Tobacco, and Firearms; Dr. N. S. Cartwright of the Royal Canadian Mounted Police, Research and Development—Chemistry; and Dr. D. M. Adams of the University of Leicester for their generous cooperation in keeping us informed of their many valuable contributions to the subject matter.

We have also made use of the Bibliography on High Pressure Research published by the High Pressure Data Center in Provo, Utah, and are grateful for their compilations.

To the people whose work we missed or left out of the book, we offer our apologies. Much of this omission was based on several contributing factors: (1) the enormous volume of publications that have accumulated during the past 24 years, (2) the effort to keep the size of the book at a reasonable level, and (3) the individual prejudices of the author in regard to choice of topics.

CHAPTER 1

# INTRODUCTION

An excellent book by Drickamer and Frank on the effects of pressure on electronic transitions in molecules appeared in 1973. To date, no related compilation on pressure effects on vibrational transitions has appeared. This book attempts to bridge this gap. Primarily it deals with the very versatile diamond anvil cell (DAC), which serves the dual purpose of generating the pressures and being transparent to infrared radiation, allowing one to make the observations of changes caused by pressure. The optical probes highlighted will deal principally with infrared and Raman scattering, although some observations in the visible region will also be presented.

This book comprises eight chapters. Chapter 1 is an introduction to the subject; Chapter 2 concerns itself with the instrumentation necessary to study vibrational transitions under pressure. The piston–cylinder and diamond anvil cell are discussed, although emphasis is on the latter cell. The spectrophotometry to obtain the spectra and the optical link between the pressure device and the spectrophotometer are discussed. Chapter 3 deals with pressure calibration and the various methods used to measure pressure in the DAC from its inception to the present.

The last portion of the book, Chapters 4–8, deals with applications. Chapters 4–6 deal with applications in basic areas of inorganic, coordination, and organic compounds. The solid state* is emphasized, and pressure-induced phase transitions are discussed. In Chapter 5 the effects of pressure on spin states and various geometries are considered. Chapter 6 deals with pressure effects on organic molecules. Chapter 7 considers more practical applications in geochemistry, conductors, forensic science, and lubricants. Chapter 8 deals with miscellaneous topics such as metallic hydrogen, metallic xenon, and CuCl.

## 1. EFFECTS OF PRESSURE

In considering the effects of high pressure on various vibrational modes in solids, one can list several general consequences that can occur:

(a) pressure phase transitions,
(b) frequency shifts induced by pressure,
(c) band shape and intensity changes,
(d) splitting of degenerate vibrations (appearance of unallowed vibrations),
(e) doubling of vibrations, and
(f) sensitivity of vibrations demonstrating expansion of molecular volume.

Other effects of pressure, such as the effects on hydrogen-bonded systems and mode softening, will be discussed in the subsequent chapters.

### a. Pressure Phase Transitions

Pressure-induced phase transitions in solids are the most dramatic changes that can occur with pressure. Often the pressure-stable phase will exist in a significantly different space group from the pressure-ambient phase. When vibrational spectroscopy is used to follow the changes, major differences in the spectra are observed. This is a consequence of the differences between the selection rules for the two phases. The DAC has been useful in

---

* A book entitled "Liquid Phase High Pressure Chemistry" has recently appeared (Isaacs, 1981).

## 1. Effects of Pressure

delineating phase diagrams involving temperature and pressure variables. A typical phase diagram is illustrated for $H_2O$ ice in Chapter 4, Fig. 26.

Phase transitions occurring in inorganic and coordination compounds are often the most dramatic ones observed. They involve significant geometrical changes and are accompanied by major changes in the vibrational spectrum as well as the visible spectrum. For a detailed discussion on these systems, see Chapters 4 and 5.

### b. Frequency Shifts

The effects of pressure on various vibrations in solids have been discussed by Sherman (Sherman and Wilkinson, 1980; Sherman, 1982a,b). Considerable interest in pressure-induced frequency shifts has developed. For molecules that are not undergoing a phase transition, the normal coordinates of the vibration remain unchanged. Since the masses also remain unchanged, any frequency shifts are due to changes in force constants. Because of these consequences, the frequency shifts become useful in determining anharmonicities existing in bonds. This topic is discussed in Chapter 4, Sections 1 and 11.

Most chemical bonds are found to be anharmonic. If they were harmonic, no pressure shifts would occur, and thus a very important aspect of molecular spectroscopy would be unavailable to scientists. Sherman and Wilkinson (1980) and Sherman (1982a,b) have considered the percentage change in force constant in terms of percentage change in interatomic distances and have found that all bond-stretching force constants are anharmonic.

Sherman has developed a model to determine the effect of pressure on force constants (and frequencies). For a molecule that can be described by a potential

$$V = A - Br^{-n} + Cr^{-m} \tag{1}$$

where $A$, $B$, $C$, $n$, and $m$ are constants and $r$ is the interatomic distance. The rate of change of the force constant $k$ with $r$ is given by

$$dk/dr = -(n + m + 3)\, k_0/r_0 \tag{2}$$

where $k_0$ and $r_0$ are the ambient pressure values of $k$ and $r$. A 6–12 Lennard–Jones potential shows a force constant change of

$$dk/k_0 = -21\, dr/r_0 \tag{3}$$

which amounts to a 21% increase in $k$ for a 1% decrease in $r$. For a 1–9 alkali–metal–halide-type potential, the increase in $k$ is 13% for a 1% decrease in $r$. Even for a strong covalent bond, a 6% increase in $k$ is found for a 1% decrease in $r$. Even the strongest bonds are anharmonic, and the weakest bonds demonstrate the highest anharmonicities.

Pressure effects for solids cause changes for both inter- and intramolecular interactions. This is illustrated by the examination of the equation

$$k_E/\delta r_E = k_I/\delta r_I \tag{4}$$

where $k_E$ is the external force constant and $k_I$ the internal force constant and $r_E$ and $r_I$ are the respective compressions. Since $k_I$ is 100 times greater and $r_I$ is 100 times smaller than $k_E$, the effect on $k_E$ will be greater than on $k_I$. The order of frequency shifts with pressure is 1–3 cm$^{-1}$ kbar$^{-1}$ for external modes. For internal bond-stretching modes, the range is 0.3–1 cm$^{-1}$ kbar$^{-1}$. Bond-bending modes are less affected and are of the order of 0.1–0.3 cm$^{-1}$ kbar$^{-1}$. It is to be understood that these are estimates for average behavior. Results may be found that are exceptions and do not follow the general trends.

### c. Band Shape and Intensity Changes

No systematic studies of band shape and intensities under pressure have been made to date. Most of the significant changes in band shape and intensity will occur when a phase changes. Fermi resonance can cause some of these changes as the pressure is increased. It would be expected that vibrations that have a significant contribution from lattice modes could manifest band-shape and intensity changes. Unquestionably, further studies in this area are needed, and these should be conducted under hydrostastic pressures.

### d. Splitting of Degenerate Vibrations

The loss of the degeneracy of E- or F-type vibrations with pressure is possible. In some cases, this may be due to a lowering of symmetry of the molecule. In other cases, in which a compound may have two or more molecules per unit cell, it is possible that the vibrations in the unit cell might couple, causing a factor-group splitting, or Davydov splitting (Davydov, 1966). Instances are possible

## 2. Pressure Units

whereby unallowed vibrations may become allowed by application of pressure.

### e. Doubling of Absorption Vibrations

In the course of various studies (Bayer and Ferraro, 1969) it was observed that a doubling of bands could occur with pressure. This may be due to a lowered site symmetry induced in the solid state by the external pressure. Alternatively, two accidentally overlapping vibrations may occur at the same frequency. These may be induced to separate because of a difference in the pressure dependencies manifested by the two vibrations or separate because of factor-group splitting as discussed in the preceding subsection.

### f. Vibrations with Expansion of Molecular Volume

The sensitivity to pressure of certain vibrations related to expansion of molecular volume has been well established. Chapter 5 discusses this topic in more detail.

For a more detailed discussion of theoretical approaches to pressure effects see the references in the bibliography at the end of this chapter.

## 2. PRESSURE UNITS

Prior to plunging into various discussions, the reader is alerted to the various units used in the pressure field. Since many readers may not be very conversant with pressure units, we have decided to deal with them at an early stage of this book. The international (SI) unit of pressure is the pascal (Pa) or newton meter$^{-2}$ (N m$^{-2}$). The interrelationships among the various units in use today are given as

$$1 \text{ bar} = 0.9869 \text{ atm} = 10^5 \text{ N m}^{-2} = 10^5 \text{ Pa} = 10^6 \text{ dyn cm}^{-2}$$

$$10^3 \text{ bar} = 1 \text{ kbar} = 10^8 \text{ N m}^{-2} = 10^8 \text{ Pa} = 1 \text{ MPa}$$

$$10^3 \text{ kbar} = 1 \text{ Mbar} = 10^{11} \text{ N m}^{-2} = 10^{11} \text{ Pa} = 100 \text{ GPa}$$

Metric conversions for various units of pressure are tabulated in Table I.

For the purpose of this text, the units kilobar and megabar will be used, since much of the literature is in these units. To provide some

## Table I
### Metric Conversions[a]

| Lb in.$^{-2}$ | Kg cm$^{-2}$ | Atm | In. Hg | Mm Hg | In. H$_2$O | Ft H$_2$O | M H$_2$O | Mbar or Mdyn m$^{-2}$ |
|---|---|---|---|---|---|---|---|---|
| 1 | 0.0703 | 0.0680 | 2.036 | 51.71 | 27.70 | 2.309 | 0.7037 | 0.0690 |
| 14.22 | 1 | 0.9678 | 28.96 | 735.5 | 394 | 32.81 | 10.00 | 0.9807 |
| 14.70 | 1.033 | 1 | 29.92 | 760 | 407 | 33.90 | 10.33 | 1.0133 |
| 0.4912 | 0.0345 | 0.0334 | 1 | 25.4 | 13.6 | 1.133 | 0.3452 | 0.03395 |
| 0.0193 | 0.00136 | 0.001315 | 0.03937 | 1 | 0.5358 | 0.04465 | 0.0106 | 0.00133 |
| 0.0361 | 0.00254 | 0.00246 | 0.0735 | 1.868 | 1 | 0.0833 | 0.0254 | 0.00249 |
| 0.4332 | 0.0305 | 0.0295 | 0.8829 | 22.42 | 12 | 1 | 0.3048 | 0.02986 |
| 1.422 | 0.1000 | 0.0967 | 2.895 | 73.55 | 39.37 | 3.281 | 1 | 0.09798 |
| 14.50 | 1.0197 | 0.9869 | 29.53 | 750 | 401.3 | 33.48 | 10.21 | 1 |
| 0.825 | 0.00439 | 0.00425 | 0.12729 | 3.2321 | 1.130 | 0.14416 | 0.04394 | 0.00431 |

[a] Normal atmospheric pressure (atm) is 101,325 N m$^{-2}$ (defined value).

***Table II***

*Naturally Occurring Pressures*

| Pressure (kbar) | Site of occurrence |
|---|---|
| 1 | Deepest part of ocean (Marianas Trench) |
| 10 | Crust–mantle interface (Mohorovicic discontinuity) |
| $1.37 \times 10^3$ | Mantle–core interface (Wichert–Gutenburg discontinuity) |
| $3.64 \times 10^3$ | Center of earth |
| $1 \times 10^8$ | Center of sun |
| $10^{10}$–$10^{14}$ | White dwarf star |
| $10^{17}$–$10^{21}$ | Neutron stars |

orientation for the extent of these units, Table II tabulates pressures occurring in various places in the universe.

## REFERENCES

Bayer, R., and Ferraro, J. R. (1969). *Inorg. Chem.* **8,** 1654.
Davydov, A. S. (1966). "Theory of Molecular Excitons." McGraw-Hill, New York.
Drickamer, H. G., and Frank, C. W. (1973). "Electron Transitions and the High Pressure Chemistry and Physics of Solids." Chapman and Hall, London.
Isaacs, N. S. (1981). "Liquid Phase High Pressure Chemistry." Wiley, New York.
Sherman, W. F. (1982a). *J. Phys. C.* **13,** 4601.
Sherman, W. F. (1982b). *J. Phys. C.* **15,** 9.
Sherman, W. F., and Wilkinson, G. R. (1980). "Advances in Infrared and Raman Spectroscopy" (R. J. H. Clark and R. E. Hester, eds.), Vol. 6, pp. 158–336. Heyden, London.

## BIBLIOGRAPHY

*General*

*Adv. High Pressure Res.* (1966–1969). **1–4.**
Bradley, R. S. (ed.) (1963). "High Pressure Physics and Chemistry," Vols. 1 and 2. Academic Press, New York.
Bradley, C. C. (ed.) (1969). "High Pressure Methods in Solid State Research." Plenum, New York.
Bundy, F. P., Hibbard, W. R., and Strong, H. M. (eds.) (1961). "Progress in Very High Pressure Research." Wiley, New York.

Drickamer, H. G. (1963). *In* "Solids under Pressure" (W. Paul and D. M. Warschauer, eds.), pp. 357–384. McGraw-Hill, New York.
Drickamer, H. G. (1974). *Angew. Chem. Int. Ed. Engl.* **13**, 39.
Drickamer, H. G., and Frank, C. W. (1973). "Electronic Transitions and the High Pressure Chemistry and Physics of Solids." Chapman and Hall, London.
Ferraro, J. R. (1979). *Coord. Chem. Rev.* **29**, 1.
Ferraro, J. R., and Basile, L. J. (1974). *Appl. Spectrosc.* **28**, 505.
Green, E. F., and Toennies, J. P. (1964). "Chemical Reactions in Shock Waves." Arnold, London.
Hamann, S. D. (1959). "Physico-Chemical Effects of Pressure." Butterworth, London.
Hamann, S. D. (1966). *In* "Advances in High Pressure Research" (R. S. Bradley, ed.), Vol. 1, p. 85. Academic Press, New York.
Jayaraman, A. (1983). *Rev. Mod. Phys.* **55**, 65.
Manghnani, M., and Akimoto, S. (eds.) (1977). "High-Pressure Research: Applications in Geophysics." Academic Press, New York.
Paul, W., and Warschauer, D. M. (eds.) (1963). "Solids under Pressure." McGraw-Hill, New York.
Sherman, W. F. (1982). *Bull. Soc. Chim. Fr.* **9-10**, 347.
Sherman, W. F., and Wilkinson, G. R. (1980). "Advances in Infrared and Raman Spectroscopy" (R. J. H. Clark and R. E. Hester, eds.), Vol. 6, pp. 158–336. Heyden, London.
Sinn, E. (1974). *Coord. Chem. Rev.* **12**, 185.
*Solid State Phys.* (1957). **6**.
*Solid State Phys.* (1962). **13**.
*Solid State Phys.* (1965). **17**.
*Solid State Phys.* (1966). **19**.
Weale, I. E. (1967). "Chemical Reactions at High Pressure." Spon, London.
Wentorf, R. H., Jr. (ed.) (1962). "Modern Very High Pressure Techniques." Butterworth, London.
Whatley, L. S., and Van Valkenburg, A. (1966). *Adv. High Pressure Res.* **1**, 327.

## Theory

Gutmann, V., and Mayer, H. (1976). "Application of the Functional Approach to Bond Variations under Pressure," Vol. 31, "Structure and Bonding," pp. 49–66. Springer-Verlag, New York.
Kumazawa, M. (1973). *High Temp.–High Pressures* **5**, 599.
Ondrechen, M. J. (1978). Intramolecular Electron Transfer Theory—A Study of Purely Electronic Effects. Ph.D. thesis, Northwestern Univ., Evanston, Illinois.
Welch, D. O., Dienes, G. J., and Paskin, A. (1978). *J. Phys. Chem.* **39**, 589.

# CHAPTER 2
# INSTRUMENTATION

## 1. INTRODUCTION

The instrumentation necessary to make infrared spectroscopic measurements at high pressures includes a high-pressure optical cell, a spectrophotometer, and the interface or optical link of the cell with the spectrophotometer. For optical experiments involving Raman scattering, no optical link is required, since the collimation and focusing properties of a laser make it unnecessary. For the megabar DAC (Mao and Bell, 1975–1976; Mao et al., 1978) the only optical techniques interfaced to date are the visible and Raman instrumentation.

## 2. HIGH-PRESSURE OPTICAL CELL

The basic components for a high-pressure cell consist of the mechanism to apply the pressure; the bearing surface; and if the cell is to be used in optical experiments, a suitable window. The window must be transparent in the electromagnetic region and in anvil-type cells must be able to withstand the load of the applied pressure. Generally, for most cells the spectroscopic requirements are opposed to the high-pressure design demands. Most cells are thus a compromise between these two opposed requirements.

Several high-pressure spectroscopic cells have been reported for use with gases, liquids, and solids. The discussion of cells in this chapter will concern itself with cells that are primarily used with condensed phases.

### a. Piston–Cylinder Cell

Perhaps the most common and most used cells are the piston–cylinder cell (e.g., Drickamer type) and the DAC cell. Fishman and Drickamer (1956) designed a modified Drickamer cell for optical studies as well as for electrical resistance and compressibility measurements (Drickamer and Balchan, 1962). Pressures in these cells is restricted to 200 kbar. The cell was further modified to be used in the high-frequency infrared region by Bradley et al. (1966). Figure 1 shows a diagram of the Drickamer cell. The Drickamer cell uses an alkali halide (NaCl preferably) as the pressure-transmitting medium. Window ports exist that are filled with the same alkali halide as the high-pressure window. Table I lists the merits and demerits of the cell and contrasts them with those of the DAC.

A number of additional variations or modifications of the Drickamer cell have appeared through the years. For details on these, see Tyner and Drickamer (1977), Nicol et al. (1972), Sherman (1966),

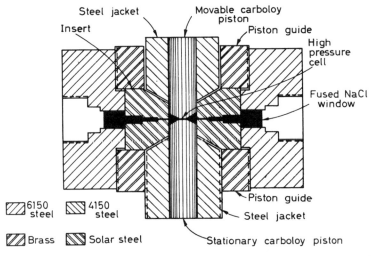

*Fig. 1.* Drickamer piston–cylinder cell. [From Drickamer and Balchan (1962).]

**Table I**

Types of Optical High-Pressure Cells

| Type of cell | Pressure attained (kbar) | Merits | Demerits |
|---|---|---|---|
| Piston–cylinder | ~180 | (1) Hydrostatic pressures<br>(2) Largest specimen volume<br>(3) Usable for liquids and solids | (1) Insufficient optical clarity to permit microscopic observation or photography of sample<br>(2) Sometimes specimen an interactant with salt matrix<br>(3) Difficult to make and use |
| DAC | 1700 | (1) Micro quantities of material<br>(2) Compact, usable with spectrophotometers and microscopes<br>(3) Usable for liquids and solids<br>(4) No matrix interference | (1) Pressure gradient in cell without gasket and transmitting fluid<br>(2) Absorption of diamonds sometimes troublesome<br>(3) Not usable with large specimen solids |

and Fitch *et al.* (1957). For a more detailed discussion of the piston–cylinder spectroscopic cell and other related cells, see Sherman and Wilkinson (1980), Whalley (1978), and Lauer (1978).

### b. Diamond Anvil Cell

Lawson and Tang (1950) at the University of Chicago first used diamonds for high-pressure studies in x-ray experiments. The cell was found to be unreliable at high pressures because of the semicylindrical holes ground into them. Pressures were limited to about 20 kbar. Jamieson improved the Lawson-and-Tang design and increased the pressure range to about 30 kbar (see Jamieson, 1957; Jamieson *et al.*, 1959).

The most versatile pressure cell used for spectroscopic studies has been the DAC, as developed by Weir *et al.* (1959), and several patents have been issued on the DAC over the course of time—1963, 1971 (on the laser heating of the DAC), and 1971 (on the Raman use of the DAC). The cell has since been modified several times and has evolved through several generations of development. Figure 2 shows the details of the original DAC. The diamonds serve as the pressure-transmitting material as well as the spectroscopic

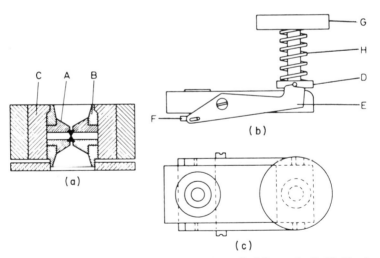

***Fig. 2.*** Diamond anvil high-pressure cell. (a) Detail of diamond cell, (b) side view, and (c) front view. A and B, parts of piston; C, hardened steel insert; D, presser plate; E, lever; G, screw; H, calibrated spring. [From Ferraro (1979).]

## 2. High-Pressure Optical Cell

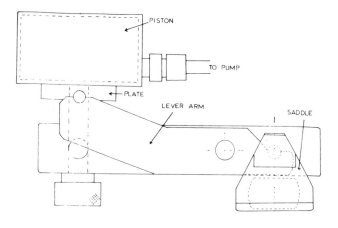

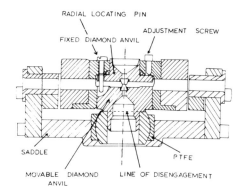

**Fig. 3.** Modified DAC. [From Adams et al. (1973).]

window. The cell can routinely be used to ~100-kbar pressure, depending on the diamond size, although higher pressures have been recorded. When gasketed, the cell can be used with liquids and solutions. The cell can also be used as a microanalytical cell, without pressure (Ferraro and Basile, 1979; Krishnan et al., 1980). A related diamond anvil cell (Bassett et al., 1967) has been used to 3000 K (Ming and Bassett, 1974) and to a low temperature of 0.03 K (Webb et al., 1976). See also Mao and Bell (1978–1979).

Piermarini and Weir (1962) were the first to modify the DAC by using a hydraulic ram. Adams et al. (1973) also used a small hydraulic ram instead of the screw in the original design. The cell is illustrated in Fig. 3. An improved version of the DAC appeared in 1973.

This was called the Waspaloy cell (Barnett *et al.*, 1973). The diamond alignment is much improved from the original instrument. One diamond is mounted in a hemisphere, and the other diamond is mounted on a plate that can be translationally positioned for axial alignment by screw adjustments. This cell has been used to 972 K and 200-kbar pressure. From this model an ultrahigh version was developed, which can be considered to be a second-generation DAC (see Piermarini and Block, 1975). The cell is illustrated in Fig. 4. It is similar in appearance and concept to the Waspaloy cell, but can be used to 550 kbar. The new cell incorporates high-strength alloy steel to obtain hardness characteristics, which are necessary for the production of the higher pressures. Alloy steel 4340 of hardness RWC 55-60 was used. For high temperatures the cell can be machined from Inconel. A lever–arm assembly with spring is still used to generate force, but Belleville washers replace the screw.

The third-generation DAC appeared in 1978 (Mao *et al.*, 1978). This is the ultrahigh pressure DAC, which has been taken to the megabar pressure region. Mao and Bell (1978) claim reaching 1.7-Mbar pressure. Figure 5 shows the cell, and Fig. 6 presents a cutaway showing the diamond anvils mounted in the cylinder of the cell (Mao and Bell, 1975–1976).

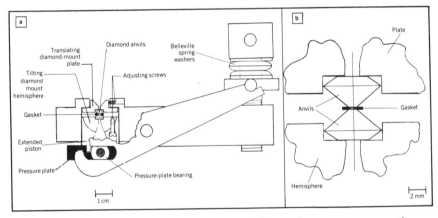

***Fig. 4.*** An ultra-high-pressure diamond cell. (a) The complete cutaway cross section shows the essential components, including the anvil supports, alignment design, lever-arm assembly, and spring-washer loading system. (b) The detail shows an enlargement of the opposed diamond anvil configuration with a metal gasket confining the sample. The cell was developed at the National Bureau of Standards. [From Block and Piermarini (1976).]

*Fig. 5.* Super-high-pressure cell for megabar use.

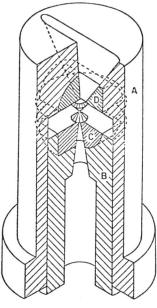

*Fig. 6.* Details of the pressure cylinder (see Mao and Bell, 1975–1976) for amplified diagrams of the entire apparatus. A, outer cylinder slotted for exit of diffracted x-ray beam (outer diameter, 2 in.); B, inner piston, tapered port for entry of primary x-ray beam; C and D, tungsten carbide half-cylinder, or "rocker," upon which the diamond surfaces are in contact.

Another cell was developed by Huber *et al.* (1977). The device has been used for single-crystal x-ray diffraction, Raman and Brillouin scattering, as well as other optical studies at high pressure.

For further discussion on types of DACs, see Jayaraman (1983).

## 3. OPTICAL WINDOWS FOR USE AT HIGH PRESSURES

In the anvil-type high-pressure cells, the window material serves the dual purpose of transmitting the pressure and being transparent to the electromagnetic radiation of interest. In the Drickamer cell this is true only in part, since the pressure transmission is not done by the window. Table II lists a number of possible window materials and their physical properties for use at high pressures. The diamond window is by far the most valuable for high-pressure studies. It is the hardest material known and is transmissive throughout most of the electromagnetic region. Figure 7 shows a single-beam spectrum of type II diamond through the DAC, as measured in a Fourier transform (FT) interferometer. Generally, several types of gem-

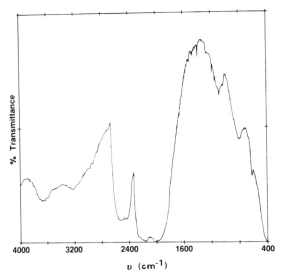

***Fig. 7.*** Expanded view of single-beam spectrum through the DAC. [From Krishnam *et al.* (1980). Reprinted from *American Laboratory*, Volume 12, Number 3, page 104, 1980, Copyright 1980 by International Science Communication, Inc.]

quality diamonds are available, and these are designated as type I and type II. Adams and Sharma (1977a) differentiate between two variations of type II (types IIa and IIb). Even within these types of diamonds, differences can be observed from diamond to diamond (Adams *et al.*, 1973; Adams and Payne, 1974; Adams and Sharma, 1977a). Type IIa is the diamond more suitable for spectroscopic work because of fewer absorptions in the infrared region. Type IIa has absorptions at ~3 and 4–5.5 μm and is transparent into the far-infrared (FIR) to ~10 cm$^{-1}$. Type IIb shows stronger absorptions at 3–5.5 μm and at ~7.7 μm. Type I shows absorptions at 3, 4–5.5, and 7–10 μm and is not suitable for the FIR.* Figure 8 shows the IR transmission spectra of type I and II diamonds from 700 to 5000 cm$^{-1}$. The infrared spectrum of sapphire is also demonstrated.

Sapphire may serve as the windows in the DAC in the mid-infrared (MID) region (2–6 μm) instead of diamond. Sapphire can also be used in the FIR, since it starts to transmit significantly at temperatures of <70 K. However, pressure limits are limited to about ~15 kbar.

For a more detailed discussion of high-pressure windows, see Sherman and Wilkinson (1980).

## 4. OPTICAL INSTRUMENTATION IN THE IR AND FOR RAMAN EXPERIMENTS

The range of wavelengths examined under high pressure with the DAC covers the gamut from the edge of the ultraviolet to the far-infrared region. In the visible and near-infrared region (overtones and combinations), energy through the DAC is sufficient that a beam-condenser mechanism may not be necessary, although in some instances it has been used (Long *et al.*, 1974). Any UV–visible spectrophotometer can be used. For the MIR both dispersive and FT–IR spectrophotometers have been utilized. In the FIR region IR instrumentation has been successful to ~500 μm (Ferraro *et al.*, 1966; Ferraro and Basile, 1980; Ferraro, 1982, unpublished data). In the infrared region (mid- and far-), beam condensers or other mechanisms of light condensation are necessary and will be discussed in a later section. Adequate and sufficient spectrophotometric instru-

---
* For a reevaluation of type I diamonds, see Wong and Klug (1983).

**Table II**

High-Pressure Optical Materials[a]

| Material | Spectral range ($\mu$m) | Refractive index[b] at $\lambda$ ($\mu$m) | Modulus of rupture[b] (psi) | Young's modulus[b] (psi) | Compressive strength (psi) | Hardness Knoop number | Solubility (g per 100 g $H_2O$) |
|---|---|---|---|---|---|---|---|
| Sodium chloride (NaCl) | 0.2–15 | 1.52 at 4<br>1.4 at 10 | | $5.8 \times 10^6$ | | 15.2–18.2 | 35.7 (0°C) |
| Lithium fluoride (LiF) | 0.11–6 | 1.35 at 4<br>1.1 at 10 | | $9.40$–$11 \times 10^6$ | | 102–113 | 0.27 (18°) |
| Irtran 1 ($MgF_2$) | 1–8 | 1.35 at 4 | 21,800 | $16.6 \times 10^6$ | 157,600 | 576 | 0.0076 (18°C) |
| Calcium fluoride ($CaF_2$) | 0.13–9 | 1.41 at 4 | | $11$–$15 \times 10^6$ | | 158 | 0.0016 (18°C) |
| Irtran 3 ($CaF_2$) | 1–10 | 1.41 at 4<br>1.34 at 8.3 | 5,300 | $14.3 \times 10^6$ | | 200 | Insoluble |
| Irtran 2 (ZnS) | 2–14 | 2.25 at 4<br>2.20 at 10 | 14,100 | $14 \times 10^6$ | 121,200 | 354 | 0.00069 (18°C) |
| Irtran 4 (ZnSe) | 0.5–20 | 2.5 at 4<br>2.4 at 10 | 6,100 | $10.3 \times 10^6$ | | 150 | Insoluble |
| Magnesium oxide (MgO) | <6.8 | 1.7 at 2.2<br>1.66 at 4.3 | | $3.6 \times 10^6$ | | 690–692 | 0.000012 |
| Irtran 5 (MgO) | 1–8 | 1.67 at 4<br>1.60 at 6 | 19,200 | $48.2 \times 10^6$ | | 640 | 0.00062 |
| Sapphire ($Al_2O_3$) | <5.5 | 1.73 at 2.2 | | $50.56 \times 10^6$ | | 1,370<br>1,525–2,000 | $9.8 \times 10^{-5}$ |

| Material | | | | |
|---|---|---|---|---|
| Diamond type I[c] | 2–4, 5–7,[d] 10–16 | 2.4173 | | |
| Diamond type II | 0.26–4; 5.5 through FIR | | 16.50 × 10⁶ | |
| Ceramic barium titanate (BaTiO₃) | <6.9 | 2.4 at 2.2, 4.3 | | |
| Calcite (CaCO₃) | 0.2–5.5 | ≃1.7 | | |
| Germanium | 1.8–2.3 | ≃4.0 | | |
| Silicon | 1–9 | 3.43 | | |
| Fused silica SiO₂ (Corning 7905, GE type 101-100, Infrasil) | 0.3–3.5 | 3.42 1.43 | | |
| NBS F158 SiO₂ | 4.5 | 1.80 | | |
| Bausch and Lomb[e] | | | | |
| RIR-2 | 4.5 | 1.75 | | |
| RIR-10, 11, 12 | 5.0 | 1.62 | | |
| RIR-20 | 5.5 | 1.82 | 7,000 | Insoluble |

[a] Taken in part from Adams and Payne (1974). Note: Whalley *et al.* (1976) and Whalley and Lavernge (1979) have reported the use of glass anvils for Raman spectra to a pressure of ~50 kbar. See also Wong and Moffatt (1983).
[b] 298 K.
[c] Diamond has the highest Debye temperature and type II diamonds demonstrate higher thermal conductivity than type I diamonds (see Ho *et al.*, 1974).
[d] UV and visible absorption and FIR depend on particular type of diamonds used here. Note that UV limitations exist for all of these materials in addition to those of diamonds.
[e] Calcium aluminate (CaO–Al₂O₃) and similar materials.

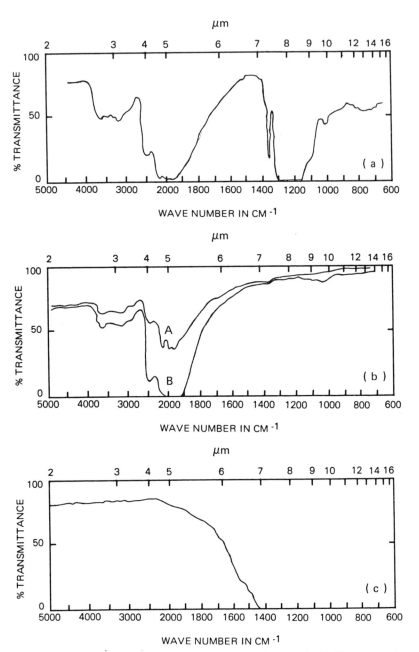

*Fig. 8.* (a) IR spectrum of typical type I diamond, 3/32 in. thick. (b) IR spectrum A, type II diamond, 3/32 in. thick; spectrum B, thin type II diamond, 1 mm thick. (c) IR spectrum of sapphire, 1 mm thick. [Reprinted with permission from Lippincott *et al.* (1961). Copyright 1961 American Chemical Society.]

## 4. Optical Instrumentation in Vibrational Experiments

mentation is presently available to accommodate the DAC for high-pressure studies using optical probes in the visible to far-infrared regions. Table III summarizes some of the optical instrumentation that has been used.

For Raman experiments at high pressures, any commercial Raman instrumentation can be used. Table IV summarizes some of the Raman instruments that have been used.

The first Raman experiment made at high pressures was performed by Gonikberg (1959), who used a piston–cylinder cell. Daniels and Hrushka (1957), Daniels (1966), and Brafman et al. (1969) used a Daniels piston–cylinder cell for Raman studies. Nicol et al. (1972) and Ferraro (1973) have used the Drickamer cell for Raman measurements at high pressure.

The first Raman experiment at high pressure with the DAC was made almost simultaneously by Postmus et al. (1968a) and Brasch and Lippincott (1968). Both 90 and 180° scattering geometries were attempted. Details of the 180° scattering are illustrated in Fig. 9 [Adams et al. (1973)]. Wong and Whalley (1974) designed diamond pressure cells for Raman studies. For Raman scattering of ultramicro samples, see Adams et al. (1977).

Some concern is necessary for the laser-stimulated fluorescence in diamonds. Adams and Payne (1974), Adams and Sharma (1977a), and Adams et al. (1973) studied the laser-stimulated fluorescence in a number of diamonds of types I, IIa, and IIb at various laser excitation wavelengths. In general, type II diamonds are to be preferred. As previously mentioned, sapphire windows can be used when high pressures are not a first-order priority. By using the 476.5-nm excita-

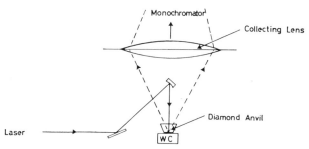

*Fig. 9.* Detail of the 180° scattering geometry in laser Raman DAC experiment. Figure not drawn to scale. [From Adams et al. (1973).]

## Table III
### Spectrophotometers Used for Optical Spectroscopy[a]

| Optical instrument | Windows | Wavelength range (μm) | Remarks | References |
|---|---|---|---|---|
| Perkin-Elmer models 421, 350; Beckman IR-4 | Diamond[b,c,d] | 2–35 | 6× beam condenser | Weir et al., 1959; Lippincott et al., 1960, 1961 |
| Perkin-Elmer models 225, 350 | Diamond | 0.27–40 | 6× beam condenser | Brasch and Jakobsen, 1965; Brasch, 1965; Jakobsen, et al., 1970 |
| Perkin-Elmer model 580 | Diamond | 2.5–55 | 4× beam condenser | Ferraro, 1982, unpublished data |
| Beckman FS-520 interferometer | Diamond | 2.5–50 | Light pipe necessary | McDevitt et al., 1967 |
| Beckman 4260 | Diamond[d] | 2–50 | With beam condenser | Ferraro and Basile, 1979 |
| | Quartz | | Cube–anvil-type cell | Bradley et al., 1966 |
| Michelson interferometer | Diamond | NIR | Cube–anvil-type cell; sample contained in NaCl; uses quartz or sapphire for lower pressure | Owen, 1966 |

| | | | | |
|---|---|---|---|---|
| RIIC-FS-720 | Diamond | 16–200 | 4× beam condenser | Ferraro et al., 1966 |
| Perkin-Elmer model 301 | Diamond | 16–200 | 6× beam condenser | Postmus et al., 1968b |
| Beckman IR-11 | Diamond | 2.5–16 | 6× beam condenser | Postmus et al., 1968b |
| Beckman IR-12 | Diamond[d] | 0.25–2.5 | 6× beam condenser | Long et al., 1974 |
| Cary 14 | Diamond | | With or without a beam condenser | |
| Digilab FTS 20 A | Diamond | 1–1000 | 6× beam condenser | Ferraro and Basile, 1980 |
| FTS-14 | Diamond | 2.5–200 | 298–673 K | R. J. Jakobsen, private communication |
| IBM 98 | Diamond | 2.5–1000 | 6× beam condenser | James et al., 1982, unpublished; Ferraro, 1982, unpublished |
| PE 421,210, Carry 14R, Spex 1700 | Diamond | 0.2–40 | To 973 K | G. J. Piermarini, private communication |
| FTS-14 | Diamond | 2.5–500 | | G. Carlson, private communication |
| Cary 14R | Diamond | 0.19–5 | 123–623 K | C. A. Angell, private communication |
| Perkin-Elmer model 225 | Diamond | | Refracting beam condenser | Adams and Sharma, 1977b |

[a] DAC can be used for liquid or solutions if a gasket is used between the anvils.
[b] DAC can be used routinely to 100 kbar if diamonds are properly aligned.
[c] Diamond type I has absorptions at 3 $\mu$m (intense); type II has absorptions at 3 $\mu$m (weak), 4–5.5 $\mu$m (intense).
[d] Sapphire windows may be used from 2 to 5 $\mu$m.

## Table IV
### Spectrophotometers Used in Raman Spectroscopy

| Optical instrument | Pressure limits (kbar) | Temperature limits (K) | Windows | Remarks | References |
|---|---|---|---|---|---|
| Spex 1401 | 100 | 77–350 | Diamond | Solids | Postmus et al., 1968b (DAC cell) |
| Spex 1401 | 100 | 77–350 | Diamond | Solids | Ferraro, 1973 (Drickamer cell) |
| Spex 1401, Cary 81 | 200 | 573 | Diamond | Solids, liquids | Brasch and Lippincott, 1968; Melveger et al., 1970 |
| Spex 1401 | 9 | 77–290 | Sapphire, quartz | Solids, gases | Lowndes, 1970 |
| Spex 1401 | 200 | To 973 | Diamond | Solids, liquids, solutions | Barnett et al., 1973 |
| Spex 1400 | 100 | 4–1300 | NaCl, diamond | Solids, liquids, solutions | Nicol et al., 1972 |
| Spex 1400 | 10 | RT | Sapphire | Solids | Brafman et al., 1969 |
| Cary 81 | 11 | 473–573 | Sapphire | Solutions, solids, liquids | Walrafen, 1968, 1973 |

| Instrument | | Temp. range | Window | Sample | Reference |
|---|---|---|---|---|---|
| Jarrell-Ash | <3 | | Sapphire | Solutions, liquids | Davis and Adams, 1971 |
| Coderg | 10 | 2–300 | Sapphire | Solids, liquids, solutions, gases | Jean-Louis and Vu, 1972 |
| Coderg | 30 | | Diamond, sapphire | Solids | Adams et al., 1973 |
| Ramanov (HG-2S) Jobin–Yvon Monochromator | >600 | | Diamond | Solids | Mammone and Sharma, 1978–1979 |
| Spex 1401 | 10 | 77–400 | Sapphire | Solids, liquids, Brillouin spectra | Peercy, 1973; Peercy and Samara, 1973; Samara and Peercy, 1973 |
| Spex 1401 | 7 | 1.4–RT | Sapphire | Solids | Durana and McTague, 1973 |
| Jarrell-Ash | 220 | 100–700 | Diamond | Solids, liquids, solutions | Wong and Whalley, 1972, 1974 |
| Spex 1405 | 3 | 223–473 | Quartz | Liquids, gases | Campbell and Jonas, 1973 |
| Cary 81 | 55 | 77–500 | Sapphire | Solids | W. F. Sherman, personal communication |

*Table V*

*Effects of Excitation Radiation When Diamond or Sapphire Is Used*

| Exciting radiation (nm) | Type II diamond | Sapphire |
|---|---|---|
| 632.8 | Entire range accessible | Useful up to 300 cm$^{-1}$ |
| 514.5 | Fluorescence too strong | Accessible to 2000 cm$^{-1}$ |
| 488.0 | Fluorescence too strong | Accessible to 3500 cm$^{-1}$ |
| 476.5 | Range accessible to 700 cm$^{-1}$ | Complete range accessible |

tion, the entire Raman region is accessible with sapphire. Correspondingly, by using the 632.8-nm excitation, the entire region is accessible with diamonds (see Table V).

## 5. OPTICAL LINK OF PRESSURE CELL TO OPTICAL CELL

A center focus on a dispersive spectrophotometer or interferometer is the best location for the DAC in the sample compartment of such instrumentation. When diamonds are used as the windows in the DAC, considerable reflection losses are incurred at each air–diamond surface. This is caused by the high refractive index of diamond ($\eta_D = 2.38$). This also restricts the angle at which light can emerge from the diamond at an air interface. The critical angle for diamond is 24° 43′, and light going through the diamond anvil and striking the interface at an angle greater than this will be internally reflected. Various other problems occur because of the small aperture in the DAC (~0.25-mm$^2$ area) and also in the Drickamer cell (cell I, ~0.028 in.; cell II, ~0.037 in.). As a consequence, a method of beam condensation of the source light must be devised.

Beam condensers condensing light 4–6× have been used for dispersive spectrophotometers and FT–IR interferometers. Figure 10 shows the beam-condenser link to the Beckman 4260 spectrophotometer. Also observed in the figure is an XYZ translator mounted to allow the DAC to be brought to the focal point quickly. In some cases, a light pipe has been used for an FT–IR interferometer (McDevitt *et al.*, 1967). Lenses have been used to cone the incoming and

## 5. Optical Link of Pressure Cell to Optical Cell

*Fig. 10.* A 6× beam condenser interfaced with a Beckman IR-4260. [From Ferraro and Basile (1979). Reprinted from *American Laboratory,* Volume 11, Number 3, page 31, 1979. Copyright 1979 by International Scientific Communications, Inc.]

outgoing light from the DAC (Adams and Sharma, 1977b; Hirschfeld, 1981). Adams and Sharma used a refracting beam condenser, which is illustrated in Fig. 11. Hirschfeld used KRS-5 lenses (hemispherical in design and coupled to the diamonds) in the MIR and effectively doubled the energy going through the diamonds when compared to energy throughput from a beam condenser. He did not attempt similar experiments for the FIR region. Presumably, hemispherical Si lenses might work. Table III lists the type of condensation used for some of the transmission optical experiments made at high pressure. Recently, the use of the DAC without any condensing technique using an FT interferometer to 100 cm$^{-1}$ has been demonstrated (Martin *et al.*, 1984).

For further reading on optical problems regarding the use of diamonds, see Adams and Sharma (1977b).

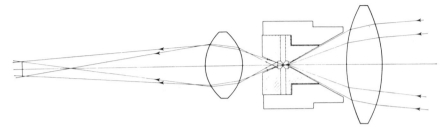

*Fig. 11.* Ray diagram of refracting beam condenser and DAC for infrared measurements. [From Adams and Sharma (1977b). Copyright The Institute of Physics, Bristol, England.]

## 6. GENERATION OF TEMPERATURES SIMULTANEOUSLY WITH PRESSURE IN THE DAC

In experiments to be described in Chapter 7, it is necessary to generate high temperatures and high pressures simultaneously in the DAC. This is particularly necessary for the study of minerals. A method using a Nd–YAG laser whereby temperatures to 3000°C and pressures to 260 kbar were generated is illustrated in Fig. 12 (Ming and Bassett, 1974). Another method is illustrated in Fig. 13 (Bell and Mao, 1975). Here again the heating is accomplished by a Nd–YAG laser. See also Hazen and Finger (1978–1979).

## 7. COMPLEMENTARY MEASUREMENTS IN THE DAC

Other measurements besides optical can be made in the DAC. The adaptability of the DAC makes this possible. It has been used in a number of diverse ways. For example, it has been used in x-ray spectroscopy (Merrill and Bassett, 1974; Bassett and Takahashi, 1974; Bassett *et al.*, 1967, 1979; Weir *et al.*, 1959; Hazen, 1977; Hazen and Finger, 1976–1977, 1978–1979, 1979–1980), Mössbauer spectroscopy (Huggins *et al.*, 1974–1975; Mao *et al.*, 1976–1977), and electrical conductivity measurements (Mao and Bell, 1973, 1977, 1981; Mao 1973a,b,c; Block and Piermarini, 1976; Walling and Ferraro, 1978; Ruoff, 1979).

## 7. Complementary Measurements in the DAC

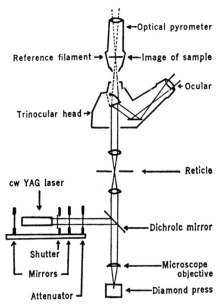

*Fig. 12.* An optical system for heating a sample under pressure in DAC using a focused laser beam. A pulsed ruby and a cw YAG laser were used. Temperature was measured by an optical pyrometer. [From Ming and Bassett (1974).]

Figure 14 illustrates a diagram of a miniature gasketed DAC for x-ray use (Merrill and Bassett, 1974; Hazen and Finger, 1979–1980).

Figure 15 shows the apparatus used in experimental resistance measurements (Mao and Bell, 1981), and Fig. 16 depicts a four-point probe method adapted for use with the DAC (Walling and Ferraro, 1978). Ruoff (1979) has designed a technique to measure resistances in small areas without any possibility of shorting. Figure 17 illustrates the nonshorting interdigitated electrode system. See also Nelson and Ruoff (1979). Figure 18 shows a schematic diagram of the DAC and the Mössbauer source. In the Mössbauer technique, a Co-57 source on the end of a palladium rod is brought close to the DAC. The gamma rays that are emitted by the source travel through the diamonds and the sample. The absorbing cation (e.g., iron) will absorb gamma rays in relation to the site, oxidation, and spin state of the metal.

One of the newer complementary techniques used with the DAC is the Brillouin scattering method. Bassett and Brody (1977) and

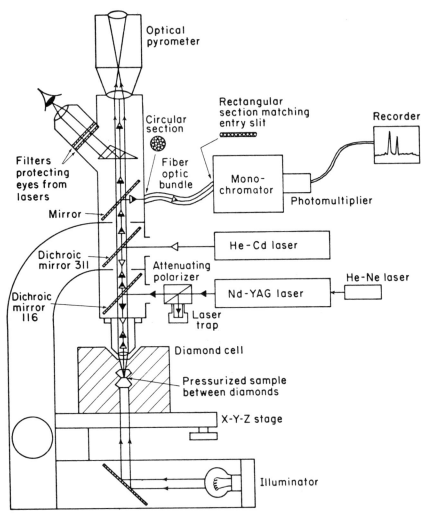

**Fig. 13.** Block diagram of laser optical system for heating experiments and pressure calibrations of the diamond-windowed high-pressure cell. ◀—, 1.06 μm Nd–YAG laser; —◀—, visible thermal emission; —◀—, 0.69 μm ruby fluorescence; —◀—, 0.44 μm He–Cd laser. Arrowheads point in the direction of the beam. The dichroic mirror 116 reflects 95% of the Nd–YAG laser at 1.06 μm and transmits 95% of the He–Cd laser at 0.44 μm, the ruby fluorescence at 0.69 μm, and the visible thermal emission of the heated sample. The dichroic mirror 311 reflects 90% of the He–Cd laser and transmits 90% of the ruby fluorescence. Each mirror can be switched in and out separately to optimize heating or pressure calibration experiments. The Nd–YAG laser is a 60-W laser pumped by two 2.5-kW krypton arc lamps. [From Bell and Mao (1975–1976).]

## 7. Complementary Measurements in the DAC

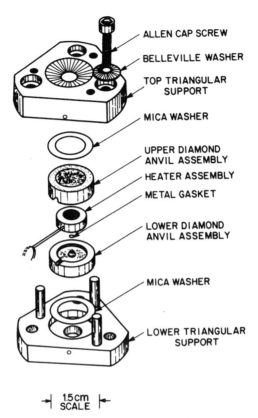

*Fig. 14.* Diagram of miniature gasketed diamond anvil pressure cell. [From Hazen and Finger (1979–1980).]

Whitfield *et al.* (1976) have designed the experiment using a laser beam and obtained elastic moduli of solids. The data are useful in the interpretation of seismic data and are helpful for gaining a better understanding of the earth's interior. The experiment is designed to scatter the laser beam from the thermal phonons in a transparent medium. On scattering, a frequency occurs due to the Doppler effect. The experimental arrangement is illustrated in Fig. 19. The frequency shift of the scattered light is Bragg reflected off the thermal phonons traversing through the solid. The Bragg reflection can be stated as

$$\lambda_{0,n} = 2\Lambda \sin \phi \qquad (1)$$

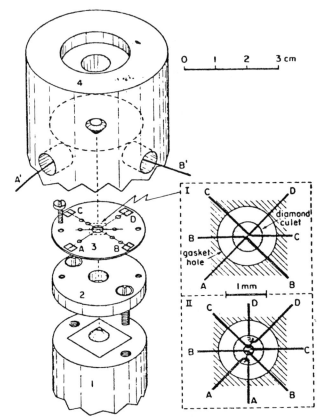

**Fig. 15.** Diagram of electrical resistance measurement apparatus with circuit for metallic conductor samples. (1) Lower piston and diamond anvil, (2) steel table to support template, (3) Mylar template with circuit cemented in place, and (4) cylinder with upper diamond anvil. (Inset I) circuit for long sample, (inset II) circuit for short sample. [From Mao and Bell (1981).]

where $\lambda_{0,n}$ is the wavelength of impinging light in the sample, $\Lambda$ the wavelength of the sound wave in the sample, and $\phi$ the angle between the incoming light and the phonon wave front. By using Snell's law,

$$n \sin \phi = (c_0 \sin \phi)/c_n = \sin \theta \tag{2}$$

where $n$ is the refractive index, $c_0$ the velocity of light in air, $c_n$ the velocity of light in the sample, and $\phi$ the angle of incidence between

## 7. Complementary Measurements in the DAC

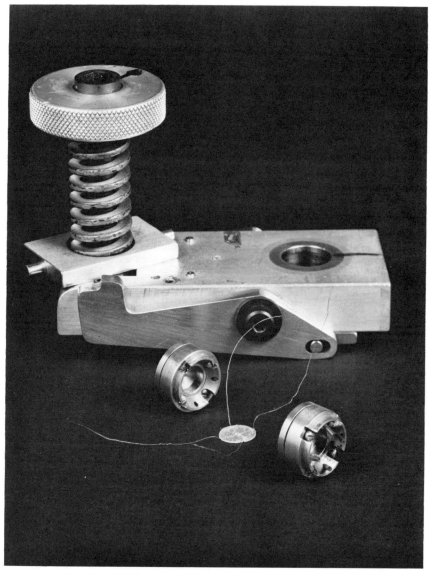

*Fig. 16.* A four-point-probe method for use in DAC. [From Walling and Ferraro (1978).]

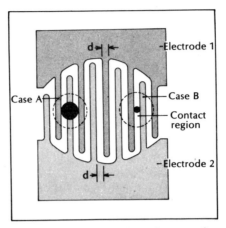

*Fig. 17.* The nonshorting interdigitated electrode system for resistance measurements in very small areas. This assembly is placed on the diamond anvil and covered with a film of nonconducting sample, and then the indentor tip is pressed against it from above. If a transition to metal occurs, the maximum size of the conducting area is determined by the spacing between the electrode fingers and their width $d$; such areas are indicated in the sketch by black circles corresponding to two possible indentor positions. Finger diameters can be made as small as a fraction of a micrometer. This system, developed in Professor Ruoff's laboratory, offers the advantage that the indentor can be positioned anywhere in the interdigitated area. The electrodes are fabricated in Cornell's National Research and Resource Facility for Submicron Structures. [From Ruoff (1979).]

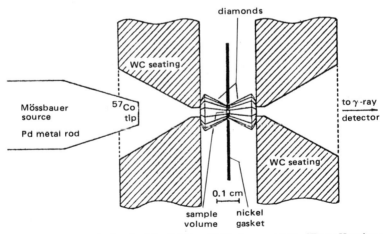

*Fig. 18.* Schematic detail of the DAC and Mössbauer source. [From Huggins *et al.* (1974–1975).]

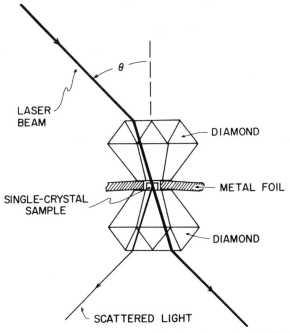

*Fig. 19.* Arrangement for Brillouin scattering experiments in a DAC. [From Bassett and Brody (1977).]

the laser beam and the outer surface of the diamond anvil. The Doppler shift is given by

$$\nu/\nu_0 = 1 \pm (2v \sin \phi)/c_n \qquad (3)$$

or

$$\pm(\nu_0 - \nu)/\nu_0 = (2v \sin \phi)/c_n \qquad (4)$$

where $\nu_0$ is the incident frequency of light, $\nu$ the frequency of light, and $v$ the velocity of the phonon. When combined with Snell's law, the relationship is expressed as

$$\pm(\nu_0 - \nu)/\nu_0 = (2v \sin \theta)/c_0 \qquad (5)$$

or

$$\pm\Delta\nu = (2v \sin \theta)/\lambda_0 \qquad (6)$$

If the angle is 45°, the equation is

$$v = (\Delta \nu\, \lambda_0)/\sqrt{2} \qquad (7)$$

which gives the relationship between frequency and velocity. The latter equation can be used to obtain elastic moduli $C_{ij}$. For example,

$$C_{ij} = \rho v^2 \qquad (8)$$

where $\rho$ = density of the material. For more details, see Shimidzu *et al.* (1982).

The DAC has recently been interfaced with energy-dispersive x-ray-diffraction (EDXRD) methods. The conventional film method of x-ray diffraction has certain limitations concerned chiefly with long exposure times (100–300 hr). These limitations make it difficult to study time-dependent phenomena, such as a rapid transformation of a metastable phase, as well as systems needing many data points for a phase diagram involving pressure, temperature, and composition. Synchrotron radiation possessing extreme spectral brilliance, well-defined polarization, and a highly collimated beam circumvents some of these difficulties. The coupling of EDXRD with high-pressure technology has been accomplished by Albritton and Margrave (1975), Bassett (1980), Mahijan *et al.* (1974), Manghnani *et al.*

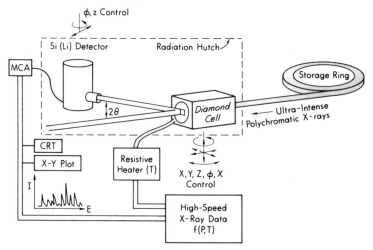

**Fig. 20.** The experimental layout for interfacing energy-dispersive x-ray diffraction (EDXRD) and a DAC using synchrotron radiation. [From Manghnani *et al.* (1981). Copyright North-Holland Publishing Company, Amsterdam, 1981.]

(1981), Skelton (1972), and Syassen and Holzapfel (1975). Figure 20 demonstrates the experimental technique for the Bassett-type and Mao–Bell DAC (Manghnani et al., 1981). One shortcoming of the technique is the need of a source of synchrotron radiation.

## REFERENCES

Albritton, L., and Margrave, J. L. (1975). *High Temp.–High Pressures* **7**, 325.
Adams, D. M., and Payne, S. J. (1974). *J. Chem. Soc. Faraday Trans.* **70**, 1959.
Adams, D. M., and Sharma, S. K. (1977a). *J. Phys. E* **10**, 680.
Adams, D. M., and Sharma, S. K. (1977b). *J. Phys. E* **10**, 838.
Adams, D. M., Payne, S. J., and Martin, K. (1973). *Appl. Spectrosc.* **27**, 377.
Adams, D. M., Sharma, S. K., and Appleby, R. (1977). *Appl Opt.* **16**, 252.
Barnett, J. D., Block, S., and Piermarini, G. J. (1973). *Rev. Sci. Instrum.* **44**, 1.
Bassett, W. A. (1980). *Phys. Earth Planet. Inter.* **23**, 337.
Bassett, W. A., and Brody, L. M. (1977). *In* "High-Pressure Research—Applications in Geophysics" (M. Manghnani and S. Akimoto, eds.), p. 519.
Bassett, W. A., and Takahashi, T. (1974). *Adv. High Pressure Res.* **4**, 165.
Bassett, W. A., Takahashi, T., and Stook, P. W. (1967). *Rev. Sci. Instrum.* **38**, 37.
Bassett, W. A., Wilburn, D. R., Hrubac, J. A., and Brody, E. M. (1979). *High-Pressure Sci. Technol. AIRAPT Conf., 6th* **2**, 75.
Bell, P. M., and Mao, H. K. (1975–1976). *Annu. Rep. Geophys. Lab., Washington, D.C.* **74**, 399.
Block, S., and Piermarini, G. J. (1976). *Phys. Today* **44**, 29.
Bradley, C. C., Gebbie, H. A., Gilbie, A. C., Ketchum, V. V., and King, J. H. (1966). *Nature (London)* **211**, 839.
Brafman, O., Mitra, S. S., Crawford, R. K., Daniels, W. B., Postmus, C., and Ferraro, J. R. (1969). *Solid State Commun.* **7**, 449.
Brasch, J. W. (1965). *J. Chem. Phys.* **43**, 3473.
Brasch, J. W., and Jakobsen, R. J. (1965). *Spectrochim. Acta* **21**, 1183.
Brasch, J. W., and Lippincott, E. R. (1968). *Chem. Phys. Lett.* **2**, 99.
Campbell, J. H., and Jonas, J. (1973). *Chem. Phys. Lett.* **18**, 441.
Daniels, W. B. (1966). *Rev. Sci. Instrum.* **37**, 1502.
Daniels, W. B., and Hruschka, A. A. (1957). *Rev. Sci. Instrum.* **28**, 1058.
Davis, A. B., and Adams, W. A. (1971). *Spectrochim Acta, Part A* **27**, 2401.
Drickamer, H. G., and Balchan, A. S. (1962). *In* "Modern Very High Pressure Techniques" (R. H. Wentorf, ed.), pp. 25–50. Butterworth, London.
Durana, S. C., and McTague, J. P. (1973). *Phys. Rev. Lett.* **31**, 990.
Ferraro, J. R. (1973). *Raman Newsletter,* No. 55, p. 17.
Ferraro, J. R. (1979). *Coord. Chem. Rev.* **29**, 1.
Ferraro, J. R. (1982). Unpublished data.
Ferraro, J. R., and Basile, L. J. (1979). *Amer. Lab. (Fairfield, Conn.)* **11**, 31.
Ferraro, J. R., and Basile, L. J. (1980). *Appl. Spectrosc.* **34**, 217.
Ferraro, J. R., Postmus, C., and Mitra, S. S. (1966). *Inorg. Nucl. Chem. Lett.* **2**, 269.
Fishman, E., and Drickamer, H. G. (1956). *Anal. Chem.* **28**, 804.

Fitch, R. A., Slykhouse, T. E., and Drickamer, H. G. (1957). *J. Opt. Soc. Am.* **47,** 1015.
Gonikberg, M. G., Stein, K. H. E., Unkholm, S. A., Opekerinov, A. A., and Aleksanias, V. T. (1959). *Opt. Spectros. (Engl. Transl.)* **6,** 166.
Hazen, R. M. (1977). *EOS, Trans. Am. Geophys. Union* **58,** 518.
Hazen, R. M., and Finger, L. W. (1976–1977). *Annu. Rep. Geophys. Lab., Washington, D.C.* **76,** 655.
Hazen, R. M., and Finger, L. W. (1978–1979). *Annu. Rep. Geophys. Lab., Washington, D.C.* **78,** 658.
Hazen, R. M., and Finger, L. W. (1979–1980). *Annu. Rep. Geophys. Lab., Washington, D.C.* **79,** 406.
Hirschfeld, T. (1981). *Pacific Conf. Chem. Spectrosc., Anaheim, California, October 19–21, 1981.*
Ho, C. Y., Powell, R. W., and Lilly, P. E. (1974). *J. Phys. Chem. Ref. Data, Suppl. 1, 3,* L-118.
Huber, G. K., Syassen, K., and Holzapfel, W. B. (1977). *Phys. Rev. B* **15,** 5123.
Huggins, F. E., Mao, H. K., and Virgo, D. (1974–1975). *Annu. Rep. Geophys. Lab., Washington, D.C.* **74,** 405.
Jakobsen, R. J., Mikawa, Y., and Brasch, J. W. (1970). *Appl. Spectrosc.* **24,** 333.
James, P. (1982). IBM Instruments Corporation, unpublished data.
Jamieson, J. C. (1957). *J. Geol.* **65,** 334.
Jamieson, J. C., Lawson, A. W., and Nachtrieb, N. D. (1959). *Rev. Sci. Instrum.* **30,** 1016.
Jayaraman, A. (1973). *Rev. Mod. Phys.* **55,** 65.
Jean-Louis, J., and Vu, H. (1972). *Rev. Phys. Appl.* **7,** 89.
Krishnan, K., Hill, S. L., and Brown, R. M. (1980). *Am. Lab. (Fairfield, Conn.)* **12,** 104.
Lauer, J. L. (1978). In "Fourier Transform Infrared Spectroscopy—Applications to Chemical Systems" (J. R. Ferraro and L. J. Basile, eds.), Vol. 1, pp. 169–213. Academic Press, New York.
Lawson, A. W., and Tang, T. Y. (1950). *Rev. Sci. Instrum.* **21,** 815.
Lippincott, E. R., Weir, C. E., Van Valkenburg, A., and Bunting, E. N. (1960). *Spectrochim. Acta* **16,** 58.
Lippincott, E. R., Welsh, F. E., and Weir, C. E. (1961). *Anal. Chem.* **33,** 137.
Long, G. J., Miles, G., and Ferraro, J. R. (1974). *Appl. Spectrosc.* **28,** 377.
Lowndes, R. P. (1970). *Phys. Rev. B* **1,** 2754.
McDevitt, N. T., Witkowski, R. E., and Fateley, W. C. (1967). *Abst., Colloq. Spectroscopium Int., 13th, June 18–24, 1967, Ottawa, Ontario, Canada.*
Mahajan, V. K., Chang, P. T., and Margrave, J. L. (1974). *High Temp.–High Pressures* **7,** 325.
Mammone, J. F., and Shirma, S. K. (1978–1979). *Annu. Rep. Geophys. Lab., Washington, D.C.* **78,** 640.
Manghnani, M. H., Ming, L. C., and Jamieson, J. C. (1980). *Nucl. Instrum. Methods* **177,** 219.
Manghnani, M. H., Skelton, E. T., Ming, L. C., Jamieson, D. C., Qadri, S., Shiferl, D., and Balogh, J. (1981) "Physics of Solids under High Pressure" (J. R. Schilling and R. N. Skelton, eds.), pp. 47–55. North-Holland Publ., Amsterdam.
Mao, H. K. (1973a). *Annu. Rep. Geophys. Lab., Washington, D.C.* **72,** 552.
Mao, H. K. (1973b). *Annu. Rep. Geophys. Lab., Washington, D.C.* **72,** 554.
Mao, H. K. (1973c). *Annu. Rep. Geophys. Lab., Washington, D.C.* **72,** 557.

Mao, H. K., and Bell, P. M. (1973). *Annu. Rep. Geophys. Lab., Washington, D.C.* **73**, 520.
Mao, H. K., and Bell, P. M. (1975–1976). *Annu. Rep. Geophys. Lab., Washington, D.C.* **75**, 824.
Mao, H. K., and Bell, P. M. (1976–1977). *Annu. Rep. Geophys. Lab., Washington, D.C.* **76**, 654.
Mao, H. K., and Bell, P. M. (1977). *In* "High-Pressure Research Applications in Geophysics" (M. H. Manghnani and S. Akimoto, eds.), pp. 493–502. Academic Press, New York.
Mao, H. K., and Bell, P. M. (1978). *Science* **200**, 1145.
Mao, H. K., and Bell, P. M. (1978–1979). *Annu. Rep. Geophys. Lab., Washington, D.C.* **78**, 659.
Mao, H. K., and Bell, P. M. (1981). *Rev. Sci. Instrum.* **52**, 615.
Mao, H. K., Virgo, D., and Bell, P. M. (1976–1977). *Annu. Rep. Geophys. Lab., Washington, D.C.* **76**, 522.
Mao, H. K., Bell, P. M., Shaner, J. W., and Steinberg, D. J. (1978). *J. Appl. Opt.* **49**, 3276.
Martin, K., Hall, L., Ferraro, J. R., and Herlinger, A. W. (1984). *Appl. Spectroscop.* [submitted].
Melveger, A. J., Brasch, J. W., and Lippincott, E. R. (1970). *Appl. Opt.* **9**, 11.
Merrill, L., and Bassett, W. A. (1974). *Rev. Sci. Instrum.* **45**, 290.
Ming, Li, and Bassett, W. A. (1974). *Rev. Sci. Instrum.* **45**, 1115.
Nelson, D. A., and Ruoff, A. L. (1979). *Phys. Rev. Lett.* **42**, 383.
Nicol, M., Ebisuzaki, Y., Ellenson, W. D., and Karim, A. (1972). *Rev. Sci. Instrum.* **43**, 1368.
Owen, N. B. (1966). *J. Sci. Instrum.* **43**, 765.
Peercy, P. S. (1973). *Phys. Rev. Lett.* **31**, 379.
Peercy, P. S., and Samara, G. A. (1973). *Phys. Rev. B* **8**, 2033.
Piermarini, G. J., and Block, S. (1975). *Phys. Sci. Instrum.* **46**, 973.
Piermarini, G. J., and Weir, C. E. (1962). *J. Res. Natl. Bur. Stand. Sect. A* **66**, 325.
Postmus, C., Ferraro, J. R., and Mitra, S. S. (1968b). *Inorg. Nucl. Chem. Lett.* **4**, 155.
Postmus, C., Maroni, V. A., and Ferraro, J. R. (1968a). *Inorg. Nucl. Chem. Lett.* **4**, 269.
Ruoff, A. L. (1979). *Cornell Quart. Eng.* **14**, 2–10.
Samara, G. A., and Peercy, P. S. (1973). *Phys. Rev. B* **7**, 1131.
Sherman, W. F. (1966). *J. Sci. Instrum.* **43**, 462.
Sherman, W. F., and Wilkinson, G. R. (1980). *Adv. Infrared Raman Spectrosc.* **6**, 158–336.
Skelton, E. F. (1972). *Rep. NRL Prog.* **1972**, 31–33.
Skelton, E. F., Liu, C. Y., and Spain, I. G. (1977). *High Temp.–High Pressures* **9**, 19.
Syassen, K., and Holzapfel, W. B. (1975). *Europhys. Conf. Abstr.* **IA**, 75.
Tyner, C. E., and Drickamer, H. G. (1977). *J. Chem. Phys.* **67**, 4103.
Walling, P. L., and Ferraro, J. R. (1978). *Rev. Sci. Instrum.* **49**, 1557.
Walrafen, G. E. (1968). *In* "Hydrogen Bonded Solvent Spectra" (A. Covington and P. Jones, eds.), p. 9. Taylor and Francis, London.
Walrafen, G. E. (1973). *J. Solution Chem.* **2**, 159.
Webb, A. W., Gubser, D. U., and Towle, C. (1976). *Rev. Sci. Instrum.* **47**, 59.
Weir, C. E., Lippincott, E. R., Van Valkenburg, A., and Bunting, E. N. (1959). *J. Res. Natl. Bur. Stand. Sec. A* **63**, 55.

Whalley, E. (1976). *Rev. Sci. Instrum.* **47**, 845.
Whalley, E. (1978). *In* "High Pressure Chemistry" (H. Kelm, ed.), pp. 127–158. D. Reidel, Dordrecht, Holland.
Whalley, E., and Lavernge, A. (1979). *Annu. Prog. Rep. Div. Chem., NRC, Canada,* p. 34.
Whalley, E., Lavernge, A., and Wong, P. T. T. (1976). *Rev. Sci. Instrum.* **47**, 845.
Whitfield, C. H., Brody, E. M., and Bassett, W. A. (1976). *Rev. Sci. Instrum.* **47**, 942.
Wong, P. T. T., and Whalley, E. (1972). *Rev. Sci. Instrum.* **43**, 935.
Wong, P. T. T., and Whalley, E. (1974). *Rev. Sci. Instrum.* **45**, 904.
Wong, P. T. T., and Klug, D. D. (1983). *Appl. Spectrosc.* **37**, 284.
Wong, P. T. T., and Moffatt, D. J. (1983). *Appl. Spectrosc.* **37**, 85.

## BIBLIOGRAPHY

### Diamond Anvil Cell

Aspandiyarov, S. B., and Vereschagin, L. F. (1970). Diamond high-pressure cell, *Sov. Phys. Dokl.* **14**, 814.
Bassett, W. A., and Kinsland, G. (1976). New applications of the diamond anvil pressure cell. I. Effect of pressure on ultimate strength in crystalline materials. *In* "Petrophysics" (R. G. Strems, ed.), pp. 71–73. Wiley, London.
Bassett, W. A., and Ming, L. C. (1976). New applications of the diamond anvil pressure cell. II. Laser treating at high pressure. *In* "Petrophysics" (R. G. Strems, ed.), pp. 365–367. Wiley, London.
Besson, J. M., and Pinceaux, J. P. (1979). Uniform stress conditions in the diamond anvil cell at 200 kbars. *Rev. Sci. Instrum.* **50**, 541.
Besson, J. M., and Pinceaux, J. P. (1974). Optics under pressure to 25 kbar. *High Temp.–High Pressures* **6**, 101.
Hayashi, S. (1979). Infrared spectra measurements using diamond high pressure cells. Trial construction of variable-pressure low-temperature cells. *Koen Yoshishu-Bunshi Kozo Sogo Toronkai,* p. 104.
Hirsch, K. R., and Holzapfel, W. B. (1981). Diamond anvil high-pressure cell for Raman spectroscopy. *Rev. Sci. Instrum.* **52**, 52.
Iwasaki, H. (1980). Diamond anvil cell for the study of alloy structures under high pressure. *Kobutsugaku Zasshi* **14**, 246.
Katahara, K. W., Manghnani, M. H., Ming, L. C., and Fisher, E. J. (1977). The bcc. transition metals under pressure: results from ultrasonic interferometry and diamond-cell experiments. *ERDA Energy Res. Abstr.* **2**(11), Abstr. No. 26968.
Kinsland, G. L. (1974). Yield strength under confining pressures to 300 kbar in the diamond anvil cell. Dissertation, Univ. Microfilms, Ann Arbor, Michigan, No. 74-22,595.
Kinsland, G. L., and Bassett, W. A. (1974). Non-isotropic strain in diamond anvil pressure cell. *EOS, Trans. Am. Geophys. Union* **55**, 415.
Kinsland, G. L., and Bassett, W. A. (1976). Modification of the diamond cell for measuring strain and the strength of materials at pressures up to 300 kbar. *Rev. Sci. Instrum.* **47**, 130.

Leute, V., and Frank, W. F. (1980). A FIR spectrometer for the investigation of polymers under high pressure. *Infrared Phys.* **20,** 327.

Liebenberg, D. H., Mills, R. L., Bronson, J. C., and Schmidt, L. C. (1978). High-pressure gases in diamond cells. *Phys. Lett A* **67,** 162.

Liebenberg, D. H., Mills, R. L., Bronson, J. C., and Schmidt, L. C. (1978). Gas loading and compression to 30 kbar with a gasketed diamond anvil cell. *Bull. Am. Phys. Soc.* **23,** 606.

Manghnani, M. H., Ming, L. C., and Jamieson, J. C. (1980). Prospects of using synchrotron radiation facilities with DAC's: High-pressure research applications in geophysics. *Nucl. Instrum. Methods* **177,** 219.

Piermarini, G. J. (1978). High-pressure physics with DAC. *Bull. Am. Phys. Soc.* **23,** 387.

Shaw, R. W., and Nicol, M. (1981). Simple low-temperature press for diamond-anvil high pressure cells. *Rev. Sci. Instrum.* **52,** 1103.

Shaw, R. W., and Nicol, M. (1975). Ultra high pressure DAC and several semiconductor phase transition pressures in relation to the fixed point pressure scale. *Rev. Sci. Instrum.* **46,** 973.

Takemura, K., Shimomura, O., Aoki, K., Asaumi, K., Minomura, S., and Tsudi, K. (1977). *Kotai Butsuri* **12,** 527.

Ves, S., and Cardona, M. (1981). A new application of the DAC: Measurements under uniaxial stress. *Solid State Commun.* **38,** 1109.

Welber, B. (1976). Optical microspectroscopic system for use with a diamond high pressure cell to 200 kbar. *Rev. Sci. Instrum.* **47,** 183.

Welber, B. (1977). Micro-optic system for reflectance measurements at pressure to 70 kbar (Application to TTF-TCNQ). *Rev. Sci. Instrum.* **48,** 395.

Wijngaarden, R. J., and Silvera, I. F. (1980). Selection of diamonds for Raman scattering in DAC's. *High-Pressure Sci. Technol. AIRAPT Conf., 7th* **1,** 157.

Yamaoka, S. (1980). A miniature diamond anvil cell and its application. *Kobutsugaki Zasshi* **14,** 241.

Yamaoka, S., and Fukunaga, O. (1979). Diamond anvil type high pressure apparatus and its application to materials science. *Seramikkuso* **14,** 601.

Yamaoka, S., Fukunaga, O., Shimomura, P., and Nakazawa, J. (1979). Versatile type miniature diamond anvil cell. *Rev. Sci. Instrum.* **50,** 1163.

## X-Ray Spectroscopy

D'Amour, H., and Holzapfel, W. B. (1980). X-ray diffraction on single crystals under pressure. *High-Pressure Sci. Technol. AIRAPT Conf., 7th* **1,** 160.

Arashi, H., Ishigame, M., and Kamiyoshi, K. (1980). Construction of a diamond anvil pressure cell for Raman scattering and x-ray diffraction measurements up to 10 GPa. *Tohoku Daigaki Kagaku Keisoku Kenkyusho Hokoku* **29,** 156.

Bassett, W. A. (1980). Synchrotron radiation. An intense x-ray source for high pressure diffraction studies. *Phys. Earth Planet. Inter.* **23,** 337.

Bassett, W. A., Hazen, R. M., and Merrill, L. (1976). New applications of the diamond anvil pressure cell. III. Single crystal diffractometer analysis at high pressure. *In* "Petrophysics" (R. G. J. Strems, ed.). Wiley, London.

Denner, W., Dietrerich, W., Schultz, M., and Holzapfel, W. B. (1978). Adaptation of a diamond anvil cell to an automatic 4-circle diffractometer for x-ray diffraction. *Rev. Sci. Instr.* **49,** 775.

Finger, L. W., and Hazen, R. M. (1978). High-pressure crystallography of minerals with the miniature diamond cell. *Acta Crystallogr. Sect. A* **34,** 344.

Finger, L. W., and King, H. (1978). A revised method of operation of the single-crystal diamond cell and refinement of the sodium chloride at 32 kbar. *Am. Mineral.* **63,** 337.

Fujii, Y., Shimomura, O., Takemura, K., Hoshino, S., and Minomura, S. (1980). The application of positive-sensitive detector to high-pressure x-ray diffraction using a diamond anvil cell. *J. Appl. Crystallogr.* **13,** 284.

Hinze, E. (1978). Optical and x-ray study of the decomposition of $Cu_2S$, $Cu_2Se$ and $Cu_2Te$ by diffusion of Cu in the pressure-gradient of a diamond cell. *Acta Crystallogr. Sect. A* **34,** 5343.

Hinze, E., Buras, B., Olsen, J. S., and Gerward, L. (1978). Energy dispension as a method for x-ray diffraction under high-pressure in a diamond anvil cell. *Acta Crystallogr. Sect. A* **34,** 5343.

Iwasaki, H., and Okada, M. (1980). X-ray diamond anvil cell and pseudo-Kossel line pattern. *High-Pressure Sci. Technol. AIRAPT Conf., 7th* **1,** 141.

Keller, R., and Holzapfel, W. B. (1977). Diamond anvil device for x-ray diffraction on single crystals under pressure up to 100 kbar. *Rev. Sci. Instrum.* **48,** 517.

Lin-Gun Liu (1980). The high pressure phase transformation of $PbO_2$: An in-situ x-ray diffraction study. *Phys. Chem. Miner.* **6,** 187.

Madon, M., Bell, P. M., Mao, H. K., and Poirier, J. P. (1980). Transmission electron diffraction and microscopy of synthetic high pressure $MgSiO_3$ phase with perovskite structure. *Geophys. Res. Lett.* **7,** 629.

Mauer, F. A., Hubbard, C. R., Peirmarini, G. J., and Block, S. (1976). Measurement of anisotropic compressibilities by a single crystal diffractometer method. *Adv. X-Ray Anal.* **18,** 437.

Merrill, L., and Bassett, W. A. (1974). Miniature diamond anvil pressure cell for single-crystal x-ray diffraction studies. *Rev. Sci. Instrum.* **45,** 290.

Merx, H., and Moussini, C. (1980). Compressibility measurements of heavy metals by energy-dispersive x-ray diffraction in the DAC. *High-Pressure Sci. Technol. AIRAPT Conf., 7th* **7,** 204.

Minomura, S., Tsuji, K., and Asaumi, K. (1980). X-ray structure analysis by DAC. *Natl. Lab. High Energy Phys.,* 272.

Newman, B. A., Pae, K. D., and Sham, T. P. (1977). High-pressure x-ray studies of polymers. II. Variation of pressure with temperature in DAC. *J. Mater. Sci.* **12,** 1697.

Newman, B. A., Sham, T. P., and Pae, K. D. (1977). High-pressure x-ray studies of polymers. I. Calibration of DAC in range 0–15 kbar. *J. Mater. Sci.* **12,** 1064.

Okada, M., and Iwasaki, H. (1980). X-ray diamond anvil cell and pseudo-Kossel line pattern. *Phys. Status Solidi A* **58,** 623.

Schifer, D. (1977). Improved beryllium diamond supports for high-pressure x-ray DAC. *High Temp.–High Pressures* **9,** 71.

Schifer, D. (1977). 50-kilobar gasket DAC for single-crystal x-ray diffractometer use

with the crystal structure of Sb up to 26 kbar as a test problem. *Rev. Sci. Instrum.* **48**, 24.
Schifer, D., Jamieson, J. C., and Lenko, J. E. (1978). 90-kbar diamond-anvil high-pressure cell for use of an automatic diffractometer. *Rev. Sci. Instrum.* **49**, 359.
Schulz, H., Denner, W., and D'Amour, H. (1977). New method of measurement using diamond high-pressure cell on a 4-circuit diffractometer. *Z. Kristallogr. Kristallogeom. Kristallphys. Kristallchem.* **146**, 163.
Schulz, H., Denner, W., and D'Amour, H. (1978). New measuring procedure for use of a gasketed anvil cell on a 4-circle diffractometer. *Acta Crystallogr. Sect. A* **34**, 345.
Sham, T. P., Newman, B. A., and Pae, K. D. (1976). High pressure x-ray studies of polymers. I. Pressure calibration in the DAC in the range of 0–15 kbars. U.S. NT15 AD Rep. No. AD-A031797, p. 17.
Shifer, D., Cromer, D. T., Mills, R. L., and Liebenberg, D. M. (1978). High pressure crystal-structure studies on nitrogen and other cryogenic gases in DAC. *Acta Crystallogr. Sect. A* **34**, 343.
Shimomura, O., Yamaoka, S., Nakazawa, N., and Fukunaga, D. (1980). X-ray intensity measurements of powdered samples in a gasketed anvil cell. *High-Pressure Sci. Technol. AIRAPT Conf., 7th* **1**, 144.
Skelton, E. F., Liu, C. Y., and Spain, I. L. (1977). Simple improvements to a diamond-anvil high-pressure cell for x-ray diffraction studies. *High Temp.–High Pressures* **9**, 19.
Skelton, E. F., Spain, I. L., Yu, S. C., and Liu, C. Y. (1977). Variable temperature pressure cell for polycrystalline x-ray studies down to 2K-application to Bi. *Rev. Sci. Instrum.* **48**, 879.
Takemura, K., Shimomura, O., Tsuji, K., and Minomura, S. (1979). Diamond-anvil pressure cell for x-ray diffraction studies with a solid-state detector or a position-sensitive proportional counter. *High Temp.–High Pressures* **11**, 311.
Takeuchi, Y. (1980). On the single crystal x-ray diffraction study at high pressure. *Kobutsugaki Zasshi* **14**, 258.
Toyoda, S., Kawamura, T., Hishima, O., and Endo, S. (1981). Film-oscillating method for x-ray diffraction with a diamond cell. *J. App. Phys.* **20**, 2201.
Wilburn, D. R., Bassett, W. A., Sto, Y., Akimoto, S. (1978). X-ray diffraction compression studies of hematite under hydrostatic isothermal conditions. *J. Geophys. Res.* **83**, 3509.

## *Electrical Measurements*

Block, S., Forman, R. A., and Piermarini, G. J. (1977). Pressure and electrical measurements in the diamond anvil cell. *In* "High Pressure Research" (M. H. Manghnani, ed.), pp. 503–508. Academic Press, New York.
Dunn, K. J., and Bundy, F. P. (1977). Electrical behavior of sulfur up to 600 kbar metallic state. *J. Chem. Phys.* **67**, 5048.
Spain, I. L., Skelton, E. F., and Rachford, F. (1980). Diamond anvil techniques for structural and electrical/magnetic measurements at low temperature. *High-Pressure Sci. Technol. AIRAPT Conf., 7th* **1**, 150.

## Brillouin Scattering

Shimidzu, H., Brody, E. M., Mao, H. K., and Bell, P. M. (1982). *Adv. in Earth Planet. Sci.* **12,** 135.

Whitfield, C. H., Brody, E. M., and Bassett, W. A. (1976). Elastic-moduli of NaCl by Brillouin-scattering at high-pressure in a DAC. *Rev. Sci. Instrum.* **47,** 949.

## Viscosity

Munro, R. G., Piermarini, G. J., and Block, S. (1979). Wall effects in a diamond-anvil pressure-cell falling sphere viscometer. *J. Appl. Phys.* **50,** 3180.

Piermarini, G. J., Forman, R. A., and Block, S. (1978). Viscosity measurements in the diamond anvil pressure cell. *Rev. Sci. Instrum.* **49,** 1061.

## Extended X-Ray Absorption Fine Structure

Shimomura, O., Fukamachi, T., Kawamura, T., Hosoya, S., Hunter, S., and Bienenstock, A. (1978). EXAFS measurements of high-pressure metallic phase of GaAs by use of a DAC. *J. Appl. Phys. Jpn.* **17,** 221.

## Explosive Materials

Brasch, J. W. (1980). Techniques for compressibility measurements on explosive materials using an opposed diamond anvil cell. *Rev. Sci. Instrum.* **51,** 1358.

CHAPTER 3

# PRESSURE CALIBRATION

## 1. INTRODUCTION

The use of the pressure variable in scientific investigations necessitates the determination of its magnitude with some degree of precision. Since our primary interest is the study of materials at high pressures and changes occurring through the use of spectroscopic probes, we shall concern ourselves with pressure calibrations in spectroscopic cells (chiefly the DAC).

The early work performed in the DAC was done on solids pressed between the diamond anvils without the use of a gasket. These conditions provided a highly nonhydrostatic environment for the solid. First Van Valkenburg (1963) and then Duecker and Lippincott (1964) demonstrated that a pressure gradient existed under these conditions that was parabolic in nature, with pressures in the center reaching $1\frac{1}{2}$ times those on the edges. Thus, any pressure measured under these conditions was basically an average pressure. Most of the early experiments (prior to 1965) were done without gasketing, and since they were nonhydrostatic, the pressures reported were average pressures. Using gaskets and incorporating pressure-transmissive fluid provide hydrostatic conditions but are applicable only to ~100 kbar. Pressures beyond this pressure are obtained essentially under nonhydrostatic or at best only quasi-hydrostatic conditions.

Under nonhydrostatic conditions, a serious question arises: Are the changes observed due to true pressure effects or to shear effects or both? Obviously, this problem would affect any method of pressure calibration. The measurement of pressure in a DAC has thus been a considerable problem to investigators in the field and continues to be so today, particularly at the higher pressures.

Van Valkenburg (1964) first used a gasket with the DAC. The use of a gasket inserted between the diamond anvils and the immersion of crystals in a fluid were major advances in the use of the DAC. This technique placed the solid of interest under virtually hydrostatic conditions, and pressure calibrations became more meaningful, especially when one could observe changes under a microscope. Various fluids have been used, but a 4 : 1 methanol–ethanol mixture has proved to be very popular (Piermarini *et al.*, 1973). Unfortunately, the use of fluids is valid only to ~100 kbar, since most liquids become solids above this pressure. The use of $H_2O$ with a methanol–ethanol mixture (16 : 3 : 1) extends the pressure range to 140 kbar (Fujishiro *et al.*, 1982); see Table I.

*Table I*

*Various Pressure Media for Hydrostatic Conditions in a DAC*

| Medium | Freezing pressure (kbar at RT) | Pressure range for hydrostatic conditions (kbar) | Reference |
|---|---|---|---|
| $CH_3OH$–EtOH (4 : 1) | 104 | ~200 | Piermarini *et al.*, 1973 |
| $H_2O$–$CH_3OH$–EtOH (16 : 3 : 1) | 145 | ~200 | Fujishiro *et al.*, 1982 |
| He | 118 | >600 | Bell and Mao, 1980–1981 |
| Ne | 47 | 160 | Fujishiro *et al.*, 1982 |
| Ar | 12 | 90 | Bell and Mao, 1980–1981 |
| Xe |  | 300 | Liebenberg, 1979 |
| $H_2$ | 57 | >600 | Mao and Bell, 1979 |
| $N_2$ | 24 | 130 | LeSar *et al.*, 1979 |

## 1. Introduction

Some debate exists as to whether simply using a gasket between the diamond anvils sans fluid provides sufficiently hydrostatic conditions to allow the worker a degree of confidence that the calibration of pressure in this situation is not an average pressure. It is probably true that the use of a gasket will reduce the pressure gradient more than when no gasket is used, but the uncertainty as to the quantity of the applied load absorbed by the gasket presents a problem.

Recent experiments involving the use of frozen gases as pressure media show much promise. LeSar (1979) reported that solid nitrogen is sufficiently plastic to be used as a pressure medium to 130 kbar. Besson and Pinceaux (1979) suggested solid helium to 150 kbar. Liebenberg (1979) recommended solid xenon to 300 kbar. Mao and Bell (1978) have used solid hydrogen to 500 kbar or higher. Mills et al. (1980) have recently suggested the use of $He^4$ and described a procedure for loading gases in a DAC. The technique is similar to that described by Mao and Bell (1978). The use of condensed gases requires special filling techniques at cryogenic temperatures.

It becomes necessary to cool the DAC for trapping and condensing the gas of choice. This is accomplished by immersing the DAC in liquid $N_2$ or $O_2$ for gases that can be condensed at or above liquid-$N_2$ temperatures. For filling at liquid-He temperatures, special cryostats are necessary to ensure the remote sealing of the fluid in the gasketed DAC. In Chapter 2 we mentioned the work of Webb et al. (1976), who used a Bassett-type cell in $He^3$–$He^4$-dilution refrigerator and obtained temperatures to 0.03 K. Shaw and Nicol (1981) designed a cryostat to reduce temperatures in a DAC to 2 K. Mao and Bell (1978) developed techniques to operate at liquid-He temperatures and for filling the gasket with hydrogen. Figure 1 shows their modified DAC for cryogenic filling. Liebenberg (1979) devised a simple technique for filling the gasket of a DAC with Xe. Figure 2 shows the design. Mills et al. (1980) described a procedure for loading a DAC with high-pressure gas. Figure 3 shows the DAC inside a pressure vessel. Besson and Pinceaux (1979) designed a high-pressure gas-loading technique for filling the gasket with He (see Fig. 4).

Table I summarizes the use of various pressure media for hydrostatic conditions in a DAC.

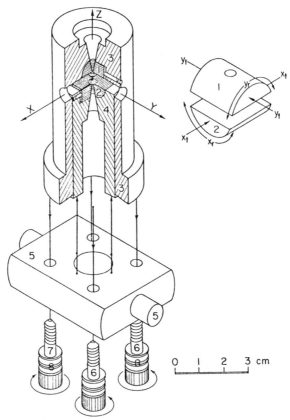

**Fig. 1.** Diamond-window, high-pressure cell for cryostatic experiments (see text). Inset to the right is an expanded view of the tungsten carbide support pieces (1) and (2), showing directions of translation and rotation for alignment of the diamond anvils. (3) Cylinder, (4) piston, and (5) thrust block. This cell as shown will operate to pressures of approximately 150 kbar by advancing the precision screws (6) and (7). (8) Belleville springs. Opposing screw directions are used to avoid torque on the remote-control system that could rupture the high-vacuum cryostat. [From Mao and Bell (1978–1979).]

## 2. METHODS OF PRESSURE CALIBRATION

Various methods of pressure calibration involving the DAC have been used. These have been based on determinations of pressure from

## 2. Methods of Pressure Calibration

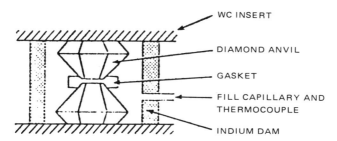

*Fig. 2.* Indium dam technique for filling and trapping liquid Xe. The diamond flats seal the gasket on applying load. [From Liebenberg (1979). Copyright North-Holland Publishing Company, Amsterdam, 1979.]

(a) an applied load,
(b) frequency dependencies of certain vibrations in various materials,
(c) phase transitions at known pressures,
(d) shifts of optical absorptions with pressure, and
(e) lattice constant dependency on pressure of NaCl as determined by means of x rays.

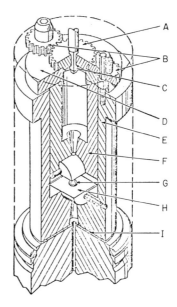

*Fig. 3.* Diamond cell inside pressure vessel. A, drive gear; B, satellite gear (two); C, square section on valve stem; D, pressure plate; E, cylinder; F, piston; G, diamond (two), copper setting not shown; H, nickel-plated surface of tungsten carbide hemicylinder (two); I, shim. [From Mills *et al.* (1980).]

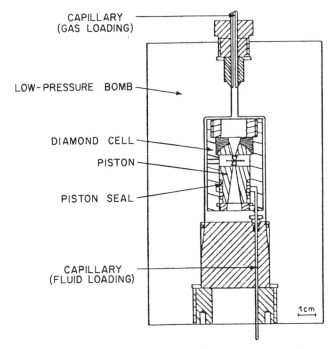

***Fig. 4.*** High-pressure gas-loading apparatus for filling DAC with He or $H_2$. [From Besson and Pinceaux (1979), *Science* **206**, 1073. Copyright 1979 by the American Association for the Advancement of Science.]

### a. Pressure Calibrated from Applied Load

At the outset of the DAC development, several workers used a method depending on the load applied to a DAC. The spring force of the DAC was obtained by a Dillon force gage, and the contact area of the diamonds was determined from photomicrographs. This allowed one to determine the pressure from the relationship $p = F/A$.

### b. Nickel Dimethylglyoxime As a Calibrant

After the advent of the use of gaskets in the DAC by Van Valkenburg (1964) [see also Weir *et al.* (1965)], more convenient and reliable methods of pressure calibration were developed. In 1968 a method was developed at the National Bureau of Standards (Davies, 1968). The method involved incorporating nickel dimethylglyoxime (DMG) in a liquid with a gasketed DAC. The pressure dependence of the visible absorption band 18,900 cm$^{-1}$ was followed versus 14 liquids with known freezing pressures. The equation for the calibration line was found to be

$$\Delta \nu = -158.9p + 1.82p^2$$

where $p$ is pressure in kilobar. Advantages of the measurement are that

(a) the measurement is hydrostatic in pressure and
(b) a relatively sharp band exists at the outset (red shift with pressure).

Disadvantages are that

(a) pressure measurements are suitable only to 100 kbar because all liquids have frozen by then,
(b) the absorption band of nickel DMG tends to lose intensity with pressure and becomes too weak to follow,
(c) temperature stability of nickel DMG is limited,
(d) nickel DMG causes spectral interference with materials under investigation, and
(e) the measuring apparatus is complex.

As a consequence of these disadvantages, the method never attained general acceptance. Other nickel complexes have also been suggested for this calibration and offer several advantages over nickel DMG (Ferraro, 1970).

## c. The Ruby Scale

In 1972 a major development occurred in the calibration of pressure in the DAC at the National Bureau of Standards. This breakthrough was successful because of the use of visual microscopic studies in the DAC. The ruby static pressure scale was developed by Forman *et al.* (1972). The technique incorporates a ruby crystal with the sample of interest and measures the pressure dependence of the sharp ruby $R_1$ line fluorescence. Actually, two fluorescences are excited ($R_1$ and $R_2$ at 692.8 and 694.2 nm), and both dependencies can be followed simultaneously. The fluorescence is induced by an Ar$^+$ or a Cd–He laser. Normally, Cr$^{3+}$ in a cubic field would show a single band due to the spin transition $^2E \leftarrow {}^4A_2$. In ruby the Cr$^{3+}$ occupies an Al$^{3+}$ site, but being of different size, it assumes a lower symmetry. Splitting of the $^2E$ level is caused by removal of degeneracy due to spin–orbit coupling and results in a separation of $R_1$ and $R_2$ of $\sim 27$ cm$^{-1}$.

The earlier work with the ruby technique was made in a gasketed DAC to 23 kbar and used known freezing pressures of several liquids that are chemically inert (Forman et al., 1972). Later the calibration was extended to $\sim 100$ kbar (Piermarini et al., 1973). Piermarini et al. (1975) calibrated the $R_1$ line at 694.2 nm to 195 kbar against the compression of NaCl by using the Decker equation of state for NaCl. The pressure dependence was found to be linear. However, nonhydrostatic conditions ensued at about 104 kbar. Adams et al. (1976) have found that even in a 4:1 methanol–ethanol mixture, nonhydrostatic conditions appear at $\sim 50$ kbar.

Eventually, the ruby pressure dependence was found to be linear to 291 kbar (Piermarini and Block, 1975) on the basis of the equation of state for NaCl. The use of fixed points involving liquids with known freezing pressures may introduce errors, because liquid-to-solid transitions create nonhydrostatic forces at the freezing pressures of the liquid. The use of fixed points involving solid–solid phase transitions may cause further errors because most of these transitions are sluggish, and determination of a pressure in these instances may not be strictly accurate.

Prior to 1975, most high-pressure experimentation with the DAC was limited to $\sim 300$ kbar. Studies with an improved DAC design and transition pressure determinations of several semiconductors, such as Si, ZnS, ZnSe, and GaP, were made by the ruby scale up to 500 kbar (Bell and Mao, 1975). These studies indicated that the revised 1970 fixed-point scale (Drickamer, 1970) and the ruby scale diverge by a factor of 2 in the 500-kbar range, with the ruby scale being lower. Advantages of the ruby scale are

(a) high intensity per unit volume—only a small ruby chip is necessary,
(b) acceptable pressure dependency (0.36 Å kbar$^{-1}$) and linearity at lower pressures,
(c) sharp linewidth (7.5 Å), and
(d) rapidity and dependability of method.

Disadvantages are

(a) presence of a doublet rather than a singlet;
(b) significant temperature coefficient (0.068 Å °C$^{-1}$) in the same direction (red) for both temperature and pressure;

## 2. Methods of Pressure Calibration

(c) thermal line broadening with pressure, causing $R_1$ and $R_2$ to overlap and limiting its use for pressure calibration to 300°C and less [heating of the ruby causes expansion along the $c$ axis, which causes symmetry to lower (from cubic), and spin–orbit coupling will increase $R_2$ and $\Delta R \simeq 29.3$ cm$^{-1}$ at 150 K to 29.7 at 293 K; both temperature and pressure cause a decrease in $R_1$ and $R_2$ frequencies, thus possibly overestimating the pressure if local heating occurs (e.g., heating in a laser)];

(d) some uncertainty in linearity of extrapolated dependency at megabar pressure and pressure broadening of both $R_1$ and $R_2$ at pressures >1 Mbar;

(e) requirement of laser excitation of fluorescence, which may be prohibitive at small laboratories or schools (at low pressures fluorescence of ruby is very weak; a highly sensitive spectrophotometer is needed to measure this fluorescence);

(f) change in $\Delta R$ under nonhydrostatic pressures (Adams *et al.*, 1976).

An illustration of the ruby scale to ~160 kbar is shown in Fig. 5.

Mao and Bell (1976) performed preliminary experiments to 1 Mbar by using ruby crystals in a gasketed ultrahigh pressure DAC. They observed the spectra shift of $R_1$ versus pressure, and their results are tabulated in Table II. In later experiments the wavelength shift of the ruby $R_1$ fluorescence line was calibrated in a Megabar DAC simultaneously versus specific measurements of Cu, Mo, Ag, and Pd (Mao and Bell, 1978, 1979). Figure 6 compares the data of Piermarini *et al.* (1975), with those of Mao *et al.* (1978). It can be observed that the ruby scale underestimates the pressure at higher pressures. The difference is within the 10% uncertainty. The equation proposed for the calibration curve was

$$p \text{ (Mbar)} = 3.808 \, ((\Delta\lambda/6942 + 1)^5 - 1)$$

where $\Delta\lambda$ is in angstroms.

Further refinements have recently been made in the high-pressure ruby calibration curve by scientists at the Geophysics Laboratory in Washington, D.C. Zou *et al.* (1981–1982) have obtained hydrostatic pressures in a modified DAC able to hold fluids and gases. Silver metal, ruby crystals, and liquefied argon were held in the sample chamber of the high-pressure cell. The specific volumes of silver were obtained by x-ray diffraction at various pressures up to

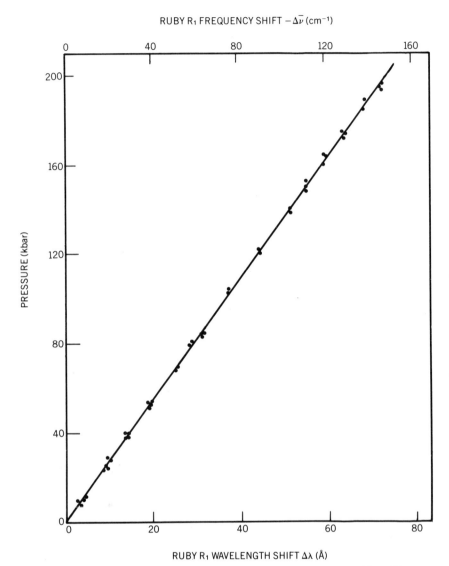

**Fig. 5.** Pressure calibration of ruby $R_1$ fluorescence line to ~160 kbar. [From Block and Piermarini (1976).]

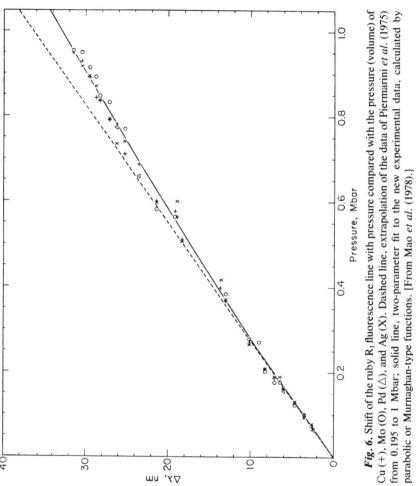

*Fig. 6.* Shift of the ruby $R_1$ fluorescence line with pressure compared with the pressure (volume) of Cu (+), Mo (O), Pd (△), and Ag (X). Dashed line, extrapolation of the data of Piermarini *et al.* (1975) from 0.195 to 1 Mbar; solid line, two-parameter fit to the new experimental data, calculated by parabolic or Murnaghan-type functions. [From Mao *et al.* (1978).]

### Table II

*Observed Spectral Shifts ($\Delta\lambda$) of the $R_1$ Line of Ruby Crystal at High Pressure*

| $\Delta\lambda$ (Å) | Pressure[a] (kbar) | $\Delta\lambda$ (Å) | Pressure[a] (kbar) |
|---|---|---|---|
| 30 | 83 | 290 | 797 |
| 75 | 206 | 310 | 823 |
| 106 | 291[b] | 320 | 880 |
| 180 | 495 | 370 | 1018 |
| 225 | 619 | | |

[a] Pressures of <291 kbar were determined from NBS calibration (Bell and Mao, 1975).

[b] $\Delta\lambda$ is the shift of the B1–B2 transition of sodium chloride.

600 kbar, and at each pressure the wavelength of the ruby $R_1$ line was measured. The largest pressure gradient was 8 kbar at a mean pressure of 600 kbar. The results are plotted in Fig. 7; also plotted is the calibration curve for nonhydrostatic conditions. In can be ob-

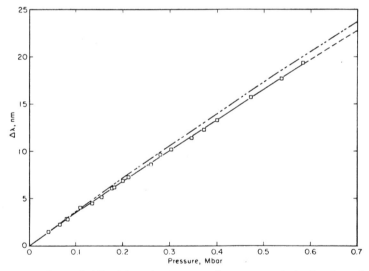

**Fig. 7.** Spectral shift of the ruby $R_1$ strong, fluorescent emission band as a function of nonhydrostatic (dashed curve) and hydrostatic (solid curve) pressure calibrated by the volume equations of state of silver (Ag). [From Zou et al. (1981–1982).]

## 2. Methods of Pressure Calibration

*Table III*

*Various Materials Whose P–V Functions Have Been Used As Secondary Pressure Scales with Reference to* NaCl *Scale*

| | |
|---|---|
| Fe | $\gamma$-$(Fe_{0.9}Mg_{0.1})SiO_4$ |
| Fe–5.2Ni | $\gamma$-$(Fe_{0.8}Mg_{0.2})SiO_4$ |
| Fe–10.3Ni | Pb |
| Fe–7.2Ni | Re |
| Fe–25Si | $\gamma$-$Co_2SiO_4$ |
| Wüstite ($Fe_{0.924}O$) | $\gamma$-$Ni_2SiO_4$ |
| Hematite ($Fe_2O_3$) | Pyrope ($Mg_3Al_2Si_3O_{12}$) |
| Magnetite ($Fe_3O_4$) | Pyrope-almandite ($Mg_{0.6}Fe_{0.3})_3Al_2Si_3O_{12}$) |
| Ilmenite (Mg, Fe)$TiO_3$ | Almandite-pyrope ($Mg_{0.2}Fe_{0.8})_3Al_2Si_3O_{12}$) |
| $\gamma$-$Fe_2SiO_4$ | Almandite ($Fe_3Al_2Si_3O_{12}$) |

served that the hydrostatic curve lies at slightly higher pressure than the nonhydrostatic curve. The pressure difference of less than 5% was reported, which is within the stated uncertainty.

Although certain questions have been raised on the validity of the ruby scale (Ruoff, 1979), it remains at present the best scale available. Until a better scale evolves, it is the pressure scale to use for DAC experimentation.

In Table III, several materials are listed whose $p$–$V$ functions can be used as secondary pressure scales with reference to the NaCl scale (see Bassett and Takahashi, 1974). For a discussion of some of the problems involving the ruby scale at megabar pressure, see Ruoff (1979). Figure 8 summarizes the present system of pressure calibration used in the DAC.

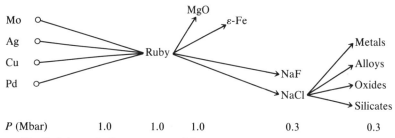

**Fig. 8.** Schematic diagram summarizing present system of calibration used in the DAC (Mao and Bell, 1979). Lines connecting two materials indicate that they were studied in simultaneous experiments. Arrows point to material calibrated. Circles indicate calibrants.

The use of the ruby technique for pressure calibrations at low temperatures presents a problem because few fixed-point transitions are known. Noack and Holzapfel (1979) calibrated the ruby scale to ~12 kbar at temperatures between 4.2 and 360 K and found that the linear scale holds at the low temperatures.

### d. Transducer Method

An alternative method of pressure calibration in the DAC can be made by the use of a pressure transducer (Sensotec, Inc., Columbus, Ohio), which can be calibrated versus known pressure transitions or versus the ruby scale. The method has been used recently by Ferraro et al. (1980). The technique is particularly appealing if no ready and repeated access to a laser is available and if one wants to use the ruby scale. The calibrations once made can be used on a continuous basis, and pressures can be easily read off the digital read out of the transducer. The calibration can be checked periodically to determine whether the calibration is still valid.

### e. Other Methods of Calibration

Because the ruby fluorescence frequency in the 0–10-kbar range is very small, it is necessary to have an extremely sensitive instrument to measure any shift of frequency. This makes it difficult for laboratories that are small and poorly equipped. Christian et al. (1975) have used methanol as an internal standard and measured the shift of the 3630 cm$^{-1}$ band in the 0–12-kbar range. They used as their primary reference the pressure data of Fishman and Drickamer (1956). Hill and Christian (1982) used anthracene dissolved in five n-alkanes and measured the anthracene band at 3750 cm$^{-1}$ as a function of pressure. These methods have not received general acceptance to date.

Several new pressure scales were reported at the ninth AIRAPT conference at SUNY, New York, July 24–28, 1983. Chopelas and Boehler (1983) suggested a MgO:V$^{2+}$ scale. The $^2E \rightarrow {}^4A_2$ fluorescence band (R line) at ~870 nm shifts with pressure and has been followed to 100 kbar. The shift was calibrated with x-ray diffraction measurements of the NaCl lattice by using a first-order Birch equation of state that was fit to accurate volume measurements (see Boehler and Kennedy, 1980). The measurements were also cross checked with the ruby fluorescence shift. The MgO:V$^{2+}$ fluorescence shifts at a rate of 0.055 ± 0.005 nm kbar$^{-1}$, over 1½ times the ruby shift. The fluorescence can be excited by a 30-mW He–Ne laser. The scale has been found to be linear to about 300°C and can be used down to 90 K. The MgO–vanadium oxide mixture can be prepared by incorporating 0.5% vanadium oxide into the MgO and heating at 1000°C in a hydrogen atmosphere.

Klug and Whalley (1983) introduced the nitrate–nitrite ion scale. Incorporating 0.3 wt. % of $NO_2^-$ and 0.12 wt. % of $NO_3^-$ into NaBr provides a mixture that can be used as an infrared pressure gage for the DAC. The antisymmetric stretching bands at 1279.0 cm$^{-1}$ ($NO_2^-$) and 1401.3 cm$^{-1}$ ($NO_3^-$) were followed with pressure up to 186 kbar relative to the $R_1$ fluorescence line of ruby. The pressure was related to the shift $\Delta\nu$ of the frequency of the $NO_2^-$ at 22°C from the zero-pressure value by the equation

$$p \text{ (kbar}^{-1}) = 2.356 \, \Delta\nu/(\text{cm}^{-1}) - 1.334(\Delta\nu/(\text{cm}^{-1})) \exp(-\Delta\nu/92 \text{ (cm}^{-1}))$$

and for the $NO_3^-$

$$p \text{ (kbar}^{-1}) = 1.775 \, \Delta\nu/(\text{cm}^{-1}) - 0.7495(\Delta\nu/(\text{cm}^{-1})) \exp(-\Delta\nu/78 \text{ (cm}^{-1}))$$

The standard deviation of pressure was reported as 1.1 and 1.4 kbar, respectively.

### f. Pressure Calibration in Piston–Cylinder Cell

The piston–cylinder cell (e.g., Drickamer cell) has also been used in spectroscopic studies to ~200 kbar. Fitch et al. (1957) calibrated the cell by measuring the rate of change of the $\nu_3$ vibration in the NCO$^-$ ion isolated in a NaCl matrix as a function of pressure to 50 kbar. Field and Sherman (1967), using the CN$^-$ ion vibration (CN stretching frequency) isolated in 12 alkali metal halides, calibrated pressures to 50 kbar in a Drickamer-type cell. Calibrations were said to be reliably extrapolated to 200 kbar. These cells have also been calibrated by measuring the bismuth and barium transition pressures or the resistance changes with pressures of various metals (e.g., Mn, Pb, Ba, Ca, and Rb) (see Drickamer and Balchan, 1962). Fox and Prengle (1969) have resorted to the use of a pressure transducer to determine pressure in this type of cell. Alternatively, some of the methods used for the DAC may be applicable (e.g., ruby technique).

For general reviews on pressure calibrations for various types of pressure cells, see Bradley (1969), Decker et al. (1973), Tsiklis (1968), Heydemann (1978), and Bassett (1979).

## REFERENCES

Adams, D. M., Appleby, R., and Sharma, S. K. (1976). *J. Phys. E* **9**, 1140.
Bassett, W. A. (1979). *Am. Rev. Earth Planet. Sci.* **7**, 357.
Bassett, W. A., and Brody, E. M. (1977). *In* "High Pressure Research—Applications in Geophysics" (M. H. Manghnani and S. Akimoto, eds.), p. 519. Academic Press, New York.
Bassett, W. A., and Takahashi, T. (1974). *Adv. High Pressure Res.* **4**, 165.
Bell, P. M., and Mao, H. K. (1980–1981). *Annu. Rep. Geophys. Lab., Washington, D.C.* **80**, 404.
Bell, P. M., and Mao, H. K. (1975). *Annu. Rep. Geophys. Lab., Washington, D.C.* **74**, 399.
Besson, J. M., and Pinceaux, J. P. (1979). *Science* **206**, 1073.
Block, S., and Piermarini, G. (1976). *Physics Today* **44**, 29.
Boehler, R., and Kennedy, G. E. (1980). *J. Phys. Chem. Solids* **41**, 517.
Bradley, C. C., ed. (1969). "High Pressure Methods in Solid State Research." Plenum, New York.
Chopelas, A., and Boehler, R. (1983). *AIRAPT Conf., 9th, SUNY, Albany, N.Y., 1983* (abstracts).
Christian, S. D., Grundes, J., and Klaboe, P. (1975). *Appl. Spectrosc.* **30**, 227.
Davies, H. W. (1968). *J. Res. Natl. Bur. Stand. Sect. A* **72**, 149.
Decker, D. L., Bassett, W. A., Merrill, L., Hall, H. T., and Barnett, J. D. (1973). "High Pressure Calibration—A Critical Review." High Pressure Data Center, Brigham Young Univ., Provo, Utah.
Drickamer, H. G. (1970). *Rev. Sci. Instrum.* **41**, 1667.
Drickamer, H. G., and Balchan, A. S. (1962). "Modern Very High Pressure Techniques" (R. N. Wentorf, ed.), p. 25. Butterworth, London.
Duecker, H. C., and Lippincott, E. R. (1964). *Science* **144**, 1119.
Ferraro, J. R. (1970). *J. Inorg. Nucl. Chem.* **6**, 823.
Ferraro, J. R., Walling, P. L., and Sherren, A. T. (1980). *Appl. Spectrosc.* **34**, 570.
Field, G. R., and Sherman, W. T. (1967). *J. Chem. Phys.* **47**, 2378.
Fishman, E., and Drickamer, H. G. (1956). *J. Chem. Phys.* **24**, 548.
Fitch, R. A., Slykhouse, T. E., and Drickamer, H. G. (1957). *J. Opt. Soc. Am.* **47**, 1015.
Forman, R. A., Piermarini, G. J., Barnett, J. D., and Block, S. (1972). *Science* **176**, 284.
Fox, J. V., and Prengle, H. W. (1969). *Appl. Spectrosc.* **23**, 157.
Fujishiro, I., Piermarini, G. J., Block, S., and Munro, R. G. (1982). *High Pressure Res. Ind. AIRAPT Conf., 8th, 1982*, p. 608.
Heydemann, P. L. M. (1978). *In* "High Pressure Chemistry" (H. Kelin, ed.), pp. 1–49. Reidel Publ., Dordrecht, Holland.

Hill, R. M., and Christian, S. D. (1982). *Appl. Spectrosc.* **36,** 302.
Klug, D. D., and Whalley, E. (1983). *AIRAPT Conf., 9th, SUNY, Albany, N.Y., 1983* (abstracts).
LeSar, R., Eleburg, S. A., Jones, L. H., Mills, R. L., Schwalbe, L. A., and Schiferl, D. (1979). *Solid State Commun.* **32,** 131.
Liebenberg, D. H. (1979). *Phys. Lett. A* **73,** 74.
Mao, H. K., and Bell, P. M. (1976). *Science* **191,** 851.
Mao, H. K., and Bell, P. M. (1978). *Science* **200,** 1145.
Mao, H. K., and Bell, P. M. (1978–1979). *Annu. Rep. Geophys. Lab., Washington, D.C.* **78,** 659.
Mao, H. K., and Bell, P. M. (1979). *Science* **203,** 1004.
Mao, H. K., Bell, P. M., Shaner, J. W., and Steinberg, D. J. (1978). *J. Appl. Phys.* **49,** 3276.
Mao, H. K., Bell, P. M., Shaner, J. W., and Steinberg, D. (1979). *High Pressure Sci. Technol. AIRAPT Conf., 6th, 1977* **1,** 739.
Mills, R. L., Liebenberg, D. H., Bronson, J. C., and Schmidt, L. C. (1980). *Rev. Sci. Instrum.* **51,** 891.
Noack, R. A., and Holzapfel, W. B. (1979). *High Pressure Sci. Technol. AIRAPT Conf., 6th, 1977* **1,** 748.
Piermarini, G. J., and Block, S. (1975). *Rev. Sci. Instrum.* **46,** 973.
Piermarini, G. J., Block, S., and Barnett, J. D. (1973). *J. Appl. Phys.* **44,** 5377.
Piermarini, G. J., Block, S., and Forman, R. A. (1975). *J. Appl. Phys.* **46,** 2774.
Ruoff, A. L. (1979). *High Pressure Sci. Technol. AIRAPT Conf., 6th, 1977* **1,** 754.
Shaw, R. W., and Nicol, M. (1981). *Rev. Sci. Instrum.* **52,** 1103.
Shumizu, H., Bassett, W. A., and Brody, E. M. (1982). *J. Appl. Phys.* **53,** 620.
Tsiklis, D. S. (1968). "Handbook of Techniques in High Pressure Research and Engineering, pp. 137–196. Plenum, New York.
Van Valkenburg, A. (1963). "High Pressure Measurements." Butterworth, London.
Van Valkenburg, A. (1964). *Ind. Diamond Rev.* **24,** *Special Suppl.,* 17.
Webb, A. W., Gubser, D. U., and Towle, C. (1976). *Rev. Sci. Instrum.* **47,** 59.
Weir, C. E., Block, S., and Piermarini, G. J. (1965). *J. Res. Natl. Bur. Stand. Sect. C* **67,** 275.
Zou, G., Bell, P. M., and Mao, H. K. (1981–1982). *Annu. Rep. Geophys. Lab., Washington, D.C.* **81,** 436.

## BIBLIOGRAPHY

Barnett, J. D., Block, S., and Piermarini, G. J. (1973). An optical fluorescence system for quantitative pressure measurement in the diamond anvil cell. *Rev. Sci. Instrum.* **44,** 1.
Bell, P. M., and Mao, H. K. (1977). Compression experiments on magnesium oxide and ruby with the diamond-window pressure cell to 1 Mbar. *In* "High Pressure Research" (M. H. Manghnani, ed.), p. 509. Academic Press, New York.
King, H. E., and Prewitt, C. T. (1980). Improved pressure calibration system using the ruby $R_1$ fluorescence. *Rev. Sci. Instrum.* **51,** 1037.

Mao, H. K., Bell, P. M., Dunn, K. J., Chrenko, R. M., and DeVries, R. C. (1979). Absolute pressure measurements and analysis of diamonds subjected to maximum static pressures of 1.3–1.7 Mbar. *Rev. Sci. Instrum* **50,** 1002.

Wunder, S. L., and Schoen, P. E. (1981). Pressure measurements at high temperature in the diamond anvil cell. *J. Appl. Phys.* **52,** 3772.

Yamaoka, S., Shumomura, O., and Fukumaga, O. (1980). Simultaneous measurements of temperature and pressure by the ruby fluorescence line. *Proc. Jpn. Acad. Ser. B* **56,** 103.

CHAPTER 4

# INORGANIC COMPOUNDS

## 1. INTRODUCTION

We are concerned in many of the compounds discussed in this chapter with ionic and pseudoionic crystals. Most of the lattice modes in these substances are located at 300 cm$^{-1}$ or lower. The successful interfacing of the DAC with a far-infrared spectrophotometer (300–25 cm$^{-1}$) was first accomplished by Ferraro et al. (1966). This has enabled one to study pressure effects on the long-wavelength optical phonons in these compounds and to use these data to determine Grüneisen parameters for them.

The relationship of the change in frequency to pressure for single cubic crystals is given by the equation

$$\gamma\chi\nu = (d\nu/dp)_T$$

where $\gamma$ is the Grüneisen parameter and $\chi$ the isothermal compressibility of the lattice mode. Most lattice modes show significant blue shifts with pressure. Since the compressibility of the solid is involved, some phonon modes may show little or no shift with pressure. For example, McDevitt et al. (1967) found little shift for zirconium and hafnium oxides up to 40 kbar. Gamma values are in the range 2–5 for the external modes in molecular crystals (Sherman and Wilkinson, 1980; Sherman, 1982). Internal modes show low $\gamma$ values of 0.02–0.1 but do increase with pressure (Sherman, 1980; Sherman and Wilkinson, 1982).

From the equation above, the $d\nu/dp$ data obtained can be used to determine the Grüneisen parameter $\gamma$. The data obtained from studies of the pressure dependency of the $k = 0$ lattice vibrations of ionic crystals when combined with data from studies of these modes as a function of temperature allow one to calculate anharmonicity taking place. It is possible to distinguish between the purely volume-dependent contribution and the contribution from anharmonic terms in the crystal Hamiltonian (Postmus et al., 1968). Figure 1 shows a plot of lg $\nu/\nu_0$ versus lg $V/V_0$ for several optical modes (Ferraro, 1979). The difference between the pressure and temperature domain can be attributed to the anharmonic contribution to the frequency shift (self-energy shift).

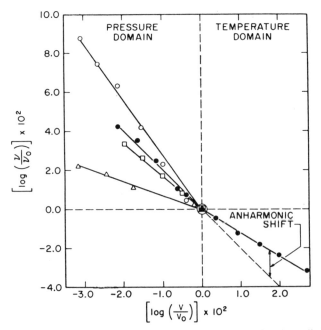

**Fig. 1.** Plot of ln $\nu/\nu_0$ versus ln $v/v_0$ for several optic modes of various alkali metal halides. ●, LiF (TO); ○, NaF (TO); △, NaF (LO); □, ZnS (TO). [From Ferraro (1971).]

## 2. ELEMENTS (NONMETALS)

### a. Nitrogen (Solid)

Solid nitrogen exists in a cubic phase $\alpha$ at low temperature and transforms to a hexagonal phase $\beta$ as the temperature is increased (Sherman and Wilkinson, 1980). A high-pressure phase $\gamma$ is formed at a pressure of 4.65 kbar and has been found to have a tetragonal structure with a space group of $D_{4h}^{14}$ with two molecules per unit cell (Medina and Daniels, 1976). Table I summarizes the selection rules and Raman and infrared activity. A fourth phase $\varepsilon$ has been reported to exist at a pressure of >49 kbar (Kobashi et al., 1982).

The far-infrared data for the $\alpha$ and $\gamma$ phases are presented in Table II (see Obriot et al., 1978, 1979). There is a noticeable shift of the two lattice modes toward higher frequency. The translational mode in the high-pressure phase $\gamma$ appears at 65.5 cm$^{-1}$.

*Table I*

*Selection Rules and Activities for $\alpha$, $\beta$, $\gamma$, and $\varepsilon$ Phases of Solid Nitrogen*

| Space group | Spectroscopic activity[a] | |
|---|---|---|
| | R | IR |
| $\alpha$, $T_h^6$ (Pa3) $Z' = 4$[b] | 5 | 2 |
| $\beta$, $D_{6h}^4$ (P6$_3$/mmc) $Z' = 2$ | 3 | 1 |
| $\gamma$, $D_{4h}^{14}$ (P4$_2$/mnn) $Z' = 2$ | 4 | 1 |
| $\varepsilon$, $O_h^3$ (Pm3n) $Z' = 8$ | 2 | 2 |

[a] Species of vibrations (active) based on space group selection rules:

$\alpha$ phase, $\Gamma_{\text{active}} = A_g + 3F_g + E_g + 2F_u$ ($A_u + E_u$ inactive; $F_u$ = acoustic),

$\beta$ phase, $\Gamma_{\text{active}} = A_{1g} + E_{1g} + E_{2g}$ ($B_{1u} + B_{2g} + E_{2u}$ inactive; $A_{2u} + E_{1u}$ = acoustic),

$\gamma$ phase, $\Gamma_{\text{active}} = A_{1g} + B_{1g} + B_{2g} + E_g + E_u$ ($B_{1u} + A_{2g}$ inactive; $A_{2u} + E_u$ = acoustic),

$\varepsilon$ phase, $\Gamma_{\text{active}} = E_g + 2F_{1g} + F_{2g} + F_{1u}$ ($A_{2g} + F_{1g} + A_{2u} + E_u + F_{2u}$ inactive; $F_{1u}$ = acoustic).

[b] $Z'$ is the number of molecules in a primitive unit cell.

### Table II

*Pressure Effects on an α Phase of Solid Nitrogen*

| Phase | Pressure (kbar)[a] | | | Mode |
|---|---|---|---|---|
| | 0.45 | 3.25 | 4.65 | |
| α | 50 cm$^{-1}$ | 61 cm$^{-1}$ | | Transitional (lattice) |
| | 70 cm$^{-1}$ | 91 cm$^{-1}$ | | Transitional (lattice) |
| γ | | | 65.5 cm$^{-1}$ | Transitional (lattice) |

[a] 4.2 K.

Medina and Daniels (1976) obtained the Raman spectra of solid nitrogen at pressures of 1–10 kbar and temperatures of 8–220 K. The high-pressure phase γ was followed as the temperature was varied, and integrated intensities were observed. Table III records these results. The two lattice modes were assigned to the $E_g$ and $B_{1g}$ vibrations in $D_{4h}^{14}$ symmetry.

### b. H₂, D₂, and Xe (Solid)

These solids are discussed in Chapter 8.

### Table III

*Observed Raman Frequencies and Relative Intensities for the γ Phase of Solid $N_2$[a]*

| Temperature (K) | $\nu$ (cm$^{-1}$) | Relative integrated intensity |
|---|---|---|
| 35 | 57.5 | 2 |
| | 95.5 | 4.5 |
| | 2329 | 1 |
| 20 | 58.2 | 5.5 |
| | 102.5 | 3 |
| | 2330 | 1 |
| 8 | 58.4 | 7 |
| | 103.6 | 2.5 |
| | 2331 | 1 |

[a] Taken in part from Medina and Daniels (1976). Note that the two lower frequencies are the lattice modes.

## c. Diamond, Silicon, and Germanium

The elemental crystals of diamond, silicon, and germanium belong to the space group $O_h^7$ ($Fd3M$) and $Z' = 2$. The active mode appears in the Raman spectrum $F_{2g}$(R). The lattice mode for diamond has been followed with pressure by Whalley et al. (1976). Figure 2 shows the blue shift of the 1332.5-cm$^{-1}$ lattice mode with pressure with $dv/dp = 0.296$ cm$^{-1}$ kbar$^{-1}$ (see Table IV).

The pressure effects on the lattice modes for Si and Ge are plotted in Fig. 3, and Table IV tabulates the $dv/dp$ data (see Buchenauer et al., 1971).

## 3. TYPE AX (II–VI AND III–V COMPOUNDS)

### Zinc Blende Family

The zinc blende family comprises a series of diatomic compounds that crystallize in a zinc blende structure. The space group is $T_d^2$ ($F^-43m$) and $Z' = 1$. Only an $F_2$ mode active in the Raman and infrared is allowed by the selection rules:

$$\Gamma_{active} = F_2(\text{IR, R})$$

The fundamental mode $F_2$ is triply degenerate, and since the zinc blende crystals are partially ionic, it corresponds at the zone center $k = 0$ of the Brillouin zone to a longitudinal optic (LO) mode, and the other to the doubly degenerate transverse optic (TO) mode. Because electromagnetic waves are transverse, they do not react with longitudinal phonons in an infinite crystal. Thus, the LO mode is usually inactive in the infrared. The TO mode is infrared active.

**Table IV**

*Lattice Mode Frequencies and Pressure Dependencies for C (Diamond), Si, and Ge*

|  | $v^a$ (cm$^{-1}$) | $dv/dp$ (cm$^{-1}$ kbar$^{-1}$) |
|---|---|---|
| C (diamond) | 1332.5 | 0.296 |
| Si | 520.2 | 0.54 |
| Ge | 300.7 | 0.46 |

[a] TO and LO modes are equal.

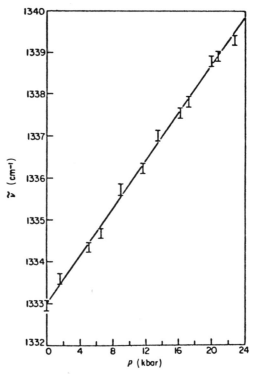

**Fig. 2.** The effect of pressure on the wave number of the fundamental Raman band of diamond. [From Whalley et al. (1976).]

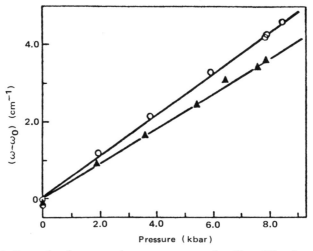

**Fig. 3.** Raman band wave number versus pressure for silicon (○) and germanium (▲). The solid lines are least-squares fits to the experimental points. [From Buchenauer et al. (1971).]

## 3. Type AX (II–VI and III–V Compounds)

**Table V**

Lattice Modes for II–VI and III–V Compounds[a]

| Compounds | $\nu_{TO}$ (cm$^{-1}$) | $\nu_{LO}$ (cm$^{-1}$) |
|---|---|---|
| *Type II–VI* | | |
| ZnO | (∥)377 | 575 |
|  | (⊥)406 | 589 |
| ZnS | 278 | 350 |
| ZnSe | 205 | 253 |
| ZnTe | 179 | 206 |
| CdS | 239 | 306 |
| CdSe | 170 | 211 |
| CdTe | 125 | 151 |
| *Type III–V* | | |
| InP | 304 | 345 |
| InAs | 219 | 243 |
| InSb | 185 | 197 |
| GaP | 367 | 403 |
| GaAs | 269 | 292 |
| GaSb | 231 | 240 |
| AlN | 667 | 916 |
| AlP | 440 | 501 |
| AlSb | 319 | 340 |
| BN | 1056 | 1304 |
| BP | 820 | 834 |

[a] Taken in part from Mitra (1969).

Table V shows locations of the TO and LO modes in II–VI and III–V compounds.

First- and second-order Raman spectra have been obtained on hexagonal ZnS (Wurtzite) at pressures to 40 kbar (Ebisuzaki and Nicol, 1972). No phase transformation to a cubic phase was observed. The Grüneisen parameters were found to be 0.99 for the LO mode and 1.81 for the TO mode.

For the TA mode the value ranged from −1.79 to 2.33 in the second-order Raman spectrum. The splitting between the TO and LO modes decreased with pressure. One- and two-phonon mode Raman spectra of GaP at pressures to 135 kbar were determined (Weinstein and Piermarini, 1975). Buchenauer et al. (1971) obtained the TO phonon mode dependency with pressure for GaAs, GaSb, and AlSb, and results are illustrated in Fig. 4.

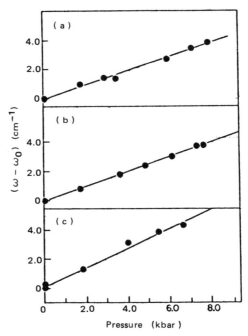

*Fig. 4.* TO phonon wave number versus pressure for (a) GaAs, (b) GaSb, and (c) AlSb. The solid lines are least-squares fits to the experimental points. [From Buchenauer *et al.* (1971).]

The pressure dependency of $k = 0$ phonon modes for some zinc blende crystals are illustrated in Fig. 5 from work by Mitra *et al.* (1969). Results of $d\nu/dp$ for both TO and LO modes are shown. These dependencies are observed to be similar for the zinc blende type crystals, with CdS and ZnO illustrating a negative pressure dependency for the degenerate $E_2$ mode.

Pressure studies of mixed crystals such as $ZnS_{1-x}Se_x$ have been made by Ferraro *et al.* (1970). These crystals are true mixed crystals in the crystallographic sense. For example, they display structures identical with that of the end members. The lattice constants of these crystals approximately follow the Vegard law in their dependency on concentration (Vegard, 1947).

These mixed crystals follow a two-mode behavior. Whether a mixed crystal $AB_{1-x}C_x$ exhibits one- or two-mode-type behavior depends almost entirely on the relative masses of the atoms A, B, and C and to a lesser extent on the strength and nature of binding. Chang

## 3. Type AX (II–VI and III–V Compounds)

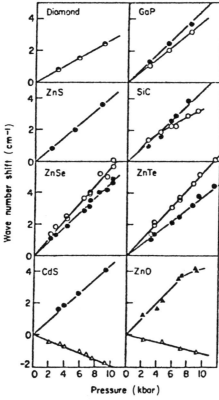

**Fig. 5.** Pressure dependency of $k = 0$ phonon wave numbers measured by Raman scattering for some crystals with zinc blende structure. [From Mitra et al. (1969).]

and Mitra (1968) have devised criteria relating the masses for the determination of the behavior type. If the relative masses are the determining factors, then an external effect such as pressure, which can modify force constants but leaves masses unaltered, should not change the one-mode-type mixed crystal to a two-mode type and vice versa. Crystals of $ZnS_{1-x}Se_x$ and $CdS_{1-x}Se_x$ display two-mode behavior to pressures to ~43 kbar.

Figure 6 shows pressure dependencies of the high- and low-frequency modes in $ZnS_{1-x}Se_x$, and Fig. 7 illustrates both modes at several mole fractions. Figure 8 presents similar data for $CdS_{1-x}Se_x$. For both mixed crystals the high-frequency mode exhibits a greater pressure dependency than does the low-frequency mode.

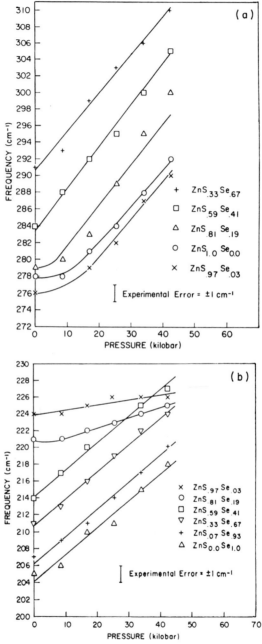

**Fig. 6.** (a) Pressure dependency of high-frequency modes in $ZnS_{1-x}Se_x$ and (b) pressure dependency of low-frequency modes in $ZnS_{1-x}Se_x$. [From Ferraro et al. (1970b).]

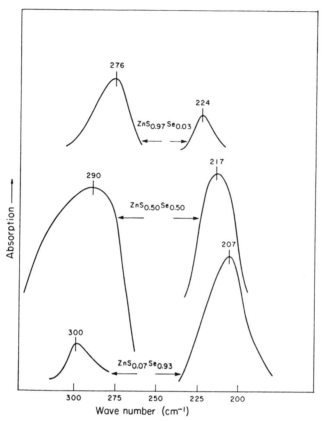

**Fig. 7.** The high-frequency and low-frequency modes of $ZnS_{1-x}Se_x$ at several mole fractions. [From Ferraro et al. (1970b).]

## 4. ALKALI METAL HALIDES (AB)

The alkali metal halides possess the NaCl-type structure ($O_h^5$, $Fm3m$) or the CsCl-type structure ($O_h^1$, $Pm3m$) with $Z' = 1$ for both types. Only a triply degenerate mode is allowed in the infrared region, but being ionic, this mode tends to split into LO and TO (doubly degenerate) modes.

$$\Gamma_{active} = F_{1u}(IR)$$

No first-order Raman spectrum is observed.

Table VI shows the TO and LO lattice mode frequencies for these salts. Normally the LO lattice mode is more difficult to study

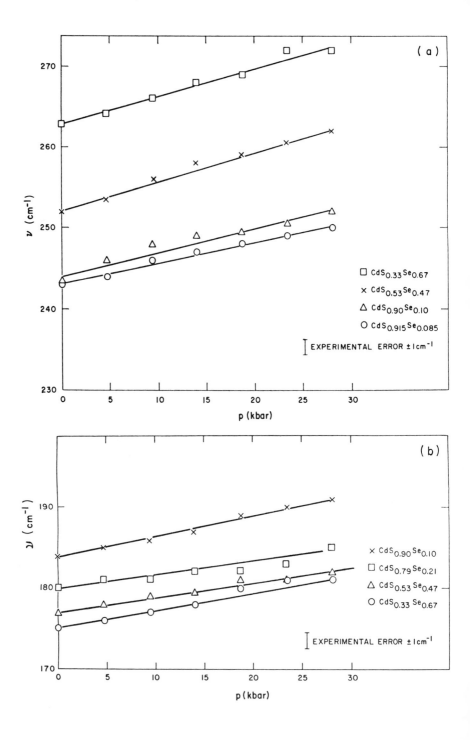

## 4. Alkali Metal Halides (AB)

**Table VI**

Lattice Vibrations for Alkali Metal Halides

| Halide | $\nu_{TO}$ (cm$^{-1}$) | $\nu_{LO}$ (cm$^{-1}$) |
|---|---|---|
| LiF | 306 | 659 |
| LiCl | 191 | 398 |
| LiBr | 159 | 325 |
| LiI | 144 | |
| NaF | 244 | 418 |
| NaCl | 164 | 264 |
| NaBr | 134 | 209 |
| NaI | 117 | 176 |
| KF | 190 | 326 |
| KCl | 142 | 214 |
| KBr | 113 | 165 |
| KI | 101 | 139 |
| RbF | 156 | 286 |
| RbCl | 116 | 173 |
| RbBr | 88 | 127 |
| RbI | 75 | 103 |
| CsF | 127 | |
| CsCl | 99 | 165 |
| CsBr | 73 | 116 |
| CsI | 62 | 85 |

in the infrared. Berreman (1963) observed the LO mode in a thin film of LiF by using oblique incident radiation. Longitudinal optic modes have been seen by similar techniques on silver halides (Bottger and Geddes, 1967). In the DAC it is possible to see the LO modes because of the highly convergent oblique radiation from the beam condenser impinging on the crystals. Generally, they appear as shoulders on the main intense TO modes and are not easily studied as functions of pressure in the infrared. With pressure the LO modes are less sensitive than the TO modes and tend to become lost in the TO mode envelope, which moves toward higher frequency with increasing pressure. In more covalent crystals, the TO mode approaches the LO mode, and in the ultimate case, for a homopolar

---

**Fig. 8.** (a) Pressure dependency of low-frequency modes in CdS$_{1-x}$Se$_x$ and (b) pressure dependency of high-frequency modes in CdS$_{1-x}$Se$_x$. [From Ferraro et al. (1970b).]

covalent crystal, the TO mode equals the LO mode (see Table IV for diamond, silicon, and germanium).

The halides of Na, K, and Rb all have the NaCl structure at atmospheric pressure. The sodium halides transform to the CsCl structure at pressures of 300 kbar. The potassium halides transform at ~20 kbar and the rubidium halides at ~5 kbar. The first far-infrared studies with pressure on these salts were made by Ferraro and co-workers (Mitra *et al.*, 1967; Postmus *et al.*, 1968; Ferraro, 1971). Figure 9 depicts the LO and TO modes for NaF as a function of pressure (Mitra *et al.*, 1967). The pressure dependencies of several alkali metal halide lattice modes are shown in Fig. 10. The TO mode was observed to drop to lower frequency in the transition to CsCl structure because of the higher coordination number in the latter structure. The vibrational frequency decreases at the transition pressure by about 10–22%. The ratio of the TO frequency in the NaCl phase with that of the CsCl phase should equal the square root

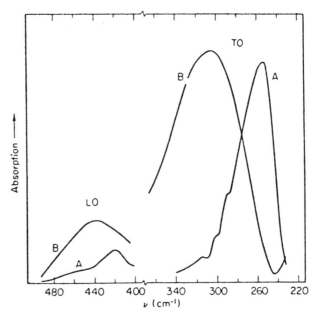

*Fig. 9.* Infrared-active lattice vibrations of NaF showing absorption due to long-wavelength LO and TO modes at two pressures: (A) 1.5 kbar and (B) 42.5 kbar. [From Ferraro (1971).]

## 4. Alkali Metal Halides (AB)

of the ratio of the coordination numbers of the two phases, e.g., $(\frac{6}{8})^{1/2}$ or 0.87. The observed ratio for NaCl is 0.88 and that of KCl is 0.92.

More recent studies at hydrostatic pressures up to 10 kbar were made by Lowndes and Rastogi (1976). Figure 11 shows the FIR of RbBr as a function of pressure. Shawyer and Sherman (1978) studied the TO mode of CsBr at elevated pressures. Figure 12 shows the pressure dependency of this salt.

Shawyer and Sherman (1982) measured the $\gamma_{TO}$ of 12 alkali metal halides under hydrostatic pressure at temperatures of 77–295 K. They obtained Grüneisen $\gamma$ parameters and pressure and temperature dependencies of $\gamma$. The data favored the hypothesis that the frequencies vary linearly with compression (fractional change in volume rather than pressure). The $\gamma$'s decrease slightly with pressure. It was previously believed that this value increased. Table VII records the $(d\nu_{TO}/dp)_T$ in reciprocal centimeters per kilobar and $(d\nu_{TO}/dV)_T$ per centimeter for the various solids studied.

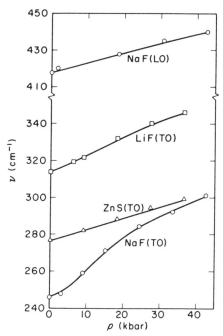

**Fig. 10.** Effect of pressure on the $k = 0$ optic mode frequencies of a few ionic crystals. [From Ferraro (1971).]

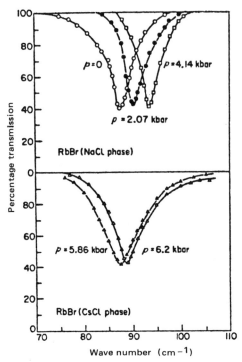

**Fig. 11.** The far-infrared absorption of RbBr as a function of pressure, with the conversion from the NaCl phase to the CsCl phase shown. [From Lowndes and Rastogi (1976).]

The mixed crystal system $KCl_{1-x}Br_x$ was investigated with pressure (Postmus *et al.*, 1970). A one-mode behavior is observed for this system. Figure 13 shows the frequency variation of the TO mode with pressure. The slopes of the TO mode versus $p$ are highest for the end members and are less for intermediate values.

Pressure dependency of the infrared phonon frequencies of $K_{1-x}Rb_xI$ was determined by Ferraro *et al.* (1971a). Figure 14 illustrates the lower frequencies and the lower $d\nu/dp$ for high-pressure phases (CsCl structure) of KI and RbI. The mixed crystal system behaves as a half-way system between the two-mode- and one-mode-type mixed crystals. Figure 15 shows the pressure dependency of the TO mode frequency for $K_{1-x}Rb_xI$.

## 4. Alkali Metal Halides (AB)

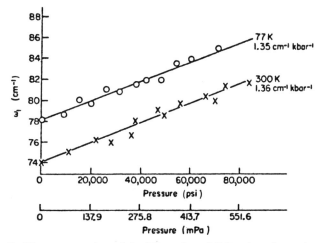

**Fig. 12.** The wave number of the TO modes of CsBr plotted as a function of pressure. All the experimentally recorded data for three runs at 300 and 77 K are shown. [From Shawyer and Sherman (1978). Reprinted with permission from *Infrared Physics*, Vol. 18. Copyright 1978, Pergamon Press, Ltd.]

*Table VII*

Slopes of $\nu_{TO}$ versus Pressure and $\nu_{TO}$ versus Compression Graphs[a]

| Sample | $(d\nu_{TO}/dp)_T$ cm$^{-1}$ kbar$^{-1}$ | | $(d\nu_{TO}/dV')_T$ cm$^{-1}$ | |
|---|---|---|---|---|
| | 77 K | 295 K | 77 K | 295 K |
| NaCl | 1.20 | 1.47 | 374 | 392 |
| NaBr | 1.52 | 1.59 | 378 | 358 |
| NaI  | 1.35 | 1.74 | 252 | 304 |
| KCl  | 1.38 | 1.77 | 284 | 307 |
| KBr  | 1.27 | 1.58 | 257 | 272 |
| KI   | 1.72 | 1.87 | 252 | 205 |
| RbCl | 1.60 | 1.73 | 280 | 296 |
| RbBr | 1.05 | 1.29 | 164 | 183 |
| RbI  | 1.37 | 1.58 | 171 | 187 |
| CsCl | 1.18 | 1.34 | 257 | 260 |
| CsBr | 1.31 | 1.39 | 242 | 243 |
| CsI  | 1.10 | 1.27 | 173 | 178 |

[a] Taken in part from Shawyer and Sherman (1982).

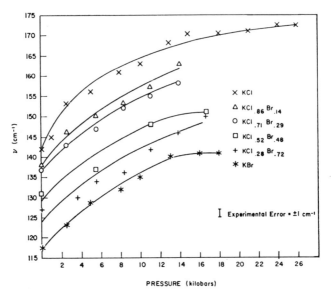

**Fig. 13.** Pressure dependency of the TO mode in $KCl_{1-x}Br_x$ system. [From Ferraro et al., Applied Optics, Vol. 9, p. 6 (1970a).]

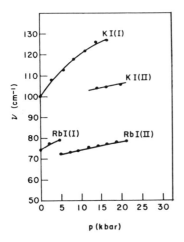

**Fig. 14.** Long-wavelength TO phonon frequency of KI and RbI as functions of pressure. Results for both the low-pressure phase (I) (NaCl structure) and the high-pressure phase (II) (CsCl structure) are shown. [From Ferraro et al. (1971a).]

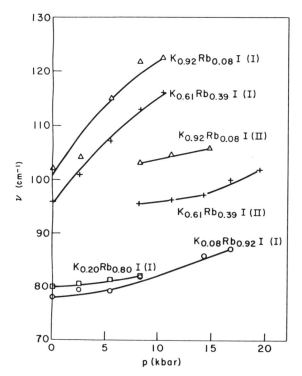

**Fig. 15.** Pressure dependency of the TO mode frequency in $K_{1-x}Rb_xI$ for four values of $x$. Phase transitions for $x = 0.08$ and $x = 0.39$ are shown. I, low-pressure phase; II, high-pressure phase. [From Ferraro et al. (1971a).]

## 5. ALKALINE-EARTH FLUORIDES

The alkaline-earth fluorides have a fluorite-type structure and belong to the $O_h^5$ ($Fm3m$) space group with $Z' = 1$. The factor group selection rules allow one infrared-active mode and one Raman-active mode:

$$\Gamma_{active} = F_{1u}(IR) + F_{2g}(R)$$

The infrared-active vibration is split (LO + TO) due to the ionic character of the salts.

Infrared and Raman studies have been made with pressure. Ferraro et al. (1971b) studied the TO mode as a function of pressure.

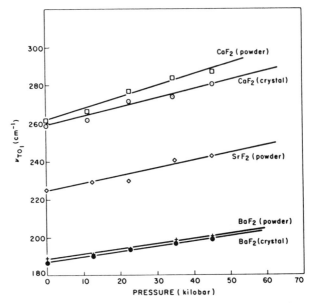

**Fig. 16.** Pressure dependency of the frequency TO mode in alkaline-earth fluorides. [From Ferraro et al. (1971a).]

**Table VIII**

*Pressure Dependencies of Several Phonon Modes of $MX_2$ Salts*

| Salt | Infrared (cm⁻¹ kbar⁻¹) | | Raman (cm⁻¹ kbar⁻¹) |
|---|---|---|---|
| | $d\nu_{LO}/dp$ | $d\nu_{TO}/dp$ | $d\nu_{LO-TO}/dp$ |
| $CaF_2$ | 0.25 | 0.55 | 0.72[a] |
| | | | 0.74[b] |
| $SrF_2$[c] | | 0.40 | 0.70 |
| $BaF_2$[c] | | 0.27 | 0.80 |
| $PbF_2$[c] | | 0.30 | — |
| $CdF_2$[c] | | 0.0 | — |

[a] Mitra (1971).
[b] Nicol et al. (1972).
[c] LO not observed.

## 6. Thallium Iodide

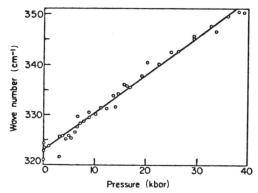

**Fig. 17.** A plot of the wave number of the Raman-active mode of $CaF_2$ versus pressure. The equation of the solid line is $\bar{\nu}$ (cm$^{-1}$) = 323 + 0.74 $p$ (kbar). [From Nicol et al. (1972).]

Figure 16 records the pressure dependencies of $CaF_2$, $SrF_2$, and $BaF_2$. Other fluorite materials such as $PbF_2$ and $PbF_2$ were also investigated. The pressure dependency of the TO mode was observed to decrease in the order $CaF_2 > SrF_2 > BaF_2 \simeq PbF_2$. Raman data by Mitra (1971) were found to illustrate greater $d\nu/dp$ dependencies for the (LO + TO) mode in these crystals. Table VIII records $d\nu/dp$ values for the salts.

Nicol et al. (1972) have studied $CaF_2$ under pressure by using Raman spectroscopy. The pressure dependency of the Raman-active mode is illustrated in Fig. 17. The $d\nu/dp$ value agrees with data by Mitra (1971).

Uniaxial pressure experiments for $CaF_2$ have been monitored by Raman spectroscopy (see Venugopalem and Ramdas, 1973).

## 6. THALLIUM IODIDE

Thallium iodide has been investigated with pressure by Brafman et al. (1969). It has a space group of $D_{2h}^{17}$ (*Cmcm*) with $Z' = 4$. The selection rules predict 24 phonon modes. At 5 kbar the salt undergoes a phase transition to a CsCl structure, and only a triply degenerate vibration becomes infrared active in this structure. Figure 18 illustrates the Raman peak height versus pressure (<5 kbar).

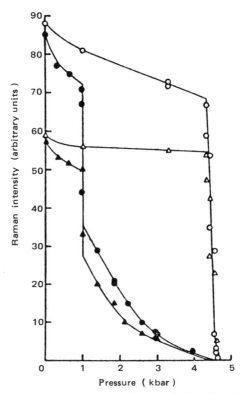

*Fig. 18.* Raman peak height versus pressure for thallium iodide. Increasing pressure is represented by open symbols, and decreasing pressure is shown by closed symbols. [From Brafman *et al.* (1969). Reprinted with permission from *Solid State Communications*, Vol. 7. Copyright 1969, Pergamon Press, Ltd.]

## 7. BIHALIDE SALTS

The halide salts of the type $RHX_2$, where R represents Na, $NH_4$, $(CH_3)_4$, and $(C_2H_5)$ and X represents F and Cl, have been investigated by infrared techniques at pressures of up to 40 kbar. Two fundamental vibrations are expected in the infrared region for a linear bihalide ion ($HX_2^-$). Hamann and Linton (1976) studied these strongly hydrogen-bonded systems and found a blue shift for the $\nu_3$ vibration, while the $\nu_2$ vibration shifts to lower energy with pressure. The $\nu_1$ frequency, unallowed in the infrared, was deduced from the behavior of combination bands and found to shift toward higher frequency. The bihalide NaHF$_2$ was found to form a new phase at 40 kbar.

## 8. AB$_2$ HALIDES*

### a. HgX$_2$

The mercuric halides have all been investigated at high pressure by using Raman and infrared spectroscopy as optical probes. Table IX shows the interrelationships of HgX$_2$ structures, and Table X tabulates selection rules for mercuric halides.

Adams and Appleby (1976, 1977a,b) have investigated the halides of Hg(II) under high pressure. The three phases (I, II, and IV) of HgCl$_2$ have been studied at pressures of up to 30 kbar. Results are tabulated in Table XI. Raman and IR spectra of phases I and IV possess a $D_{2h}^{16}$ structure with $Z' = 4$. The transition from phase I to phase IV involved a second-order transition characterized by molecular orientation with no change in space group occurring. The IV–II

*Table IX*

*Interrelationships of HgX$_2$ Structures[a]*

| | | | | |
|---|---|---|---|---|
| HgCl$_2$ | Phases I, IV (4.6 kbar) $D_{2h}^{16}$, $Z' = 4$ | | Phase II (193 kbar) $T_h^6$, $Z' = 4$ | |
| HgBr$_2$ | Phase I $C_{2v}^{12}$, $Z' = 2$ | Phase II (6.3 kbar) (like phase I?) | Phase III (23 kbar) $C_{2h}$ structure (?) | Phase IV (36 kbar) CdI$_2$ structure |
| HgI$_2$ | Red phase III $D_{4h}^{15}$, $Z' = 2$ | Phase II—yellow (127°C) $C_{2v}^{12}$, $Z' = 2$ Yellow phase (13 kbar) (like phase III—HgBr$_2$) | | Phase VI (>75 kbar) CdI$_2$ structure |

[a] $Z'$ = primitive unit cell.

* Compounds such as HgX$_2$, CdX$_2$, and PbFX.

## Table X
### Tabulation of Selection Rules for Mercuric Halides

| Phase | Space group | $Z'$ | Selection rules[a] |
|---|---|---|---|
| $HgCl_2$ | | | |
| I | $D_{2h}^{16}$ ($Pmma$) | 4 | $\Gamma = 6A_g + 3B_{1g} + 6B_{2g} + 3B_{3g}$ (R) (R) (R) (R) $+ \underline{3A_u} + 5B_{1u} + 2B_{1u} + 5B_{3u}$ (IR) (IR) (IR) |
| II | $T_h^6$ ($Pa3$) | 4 | $\Gamma = A_g + E_g + 3F_g + \underline{2A_u}$ (R) (R) R $+ 2E_u + 5F_u$ (IR) |
| IV | $D_{2h}^{16}$ ($Pmma$) | 4 | |
| $HgBr_2$ | | | |
| I | $C_{2v}^{12}$ | 2 | $\Gamma = 4A_1 + 4A_2 + 3B_1 + 4B_2$ |
| II | $C_{2v}^{12}$ (probably) | | (R, IR) (R) (R, IR) (R, IR) |
| III | $C_{2h}$ (deduced from pressure studies) | | |
| IV | $CdI_2$ (deduced from pressure studies) | | |
| $HgI_2$ | | | |
| Red | $D_{4h}^{15}$ ($P4_2/nmc$) | 2 | $\Gamma = A_{1g} + 2B_{1g} + 3E_g + A_{2u}$ (R) (R) (R) (IR) $+ B_{2u} + 2E_u$ (IR) |
| Yellow (127°C) | $C_{2v}^{12}$ ($Cmc2_1$) | 2 | $\Gamma = 4A_1 + 4A_2 + 3B_2 + 4B_3$ (R, IR) (R) (R, IR) (R, IR) |

[a] Underlined indicates inactive.

transition at 19.3 kbar involves a change in space group postulated as $T_h^6$ with $Z' = 4$.

All four phases (I–IV) of $HgBr_2$ have been investigated by Raman and IR methods up to 50 kbar by Adams and Appleby (1977b). Phases I and II were considered to have similar structures. Phase III appears to possess a $C_{2h}$ space group, and phase IV probably has a $CdI_2$ structure. These conclusions were based on the spectroscopic results. Table XII tabulates the Raman frequencies at various pressures. Figure 19 illustrates the Raman spectrum of phases I–IV.

## Table XI
*Raman and IR Results for Phases I, IV, and II in the HgCl$_2$ System*

| Phase | Raman (cm$^{-1}$) | IR (cm$^{-1}$) |
|---|---|---|
| I (ambient pressure) | 383 (vvw) | 370 (vs) |
| | 315 (s, sp) | 330 (vw) |
| | 167 (vvw) | 310 (w) |
| | 126 (w) | 100 (vs) |
| | 74 (m) | 75 (sh) |
| | 43 (vw) | |
| | 23 (m) | |
| | 18 (s) | |
| IV (8.5 kbar) | 386 (vvw) | 370 (vs) |
| | 316 (s, sp) | 330 (w) |
| | 170 (vvw) | 310 (w) |
| | 144 (w) | 100 (vs) |
| | 133 (w) | 77 (sh) |
| | 77 (m) | |
| | 51 (w, sh) | |
| | 41 (vvw) | |
| | 21 (m) | |
| II (30 kbar) | 312 (vs) | 369 (vs) |
| | 178 (s) | 100 (vs) |
| | | 72 (vs) |

## Table XII
*Raman Frequencies (Cm$^{-1}$) for HgBr$_2$ at Various Pressures (RT)*

| Pressure (kbar)/Phase | | | | | | |
|---|---|---|---|---|---|---|
| 0/I | 6.3/II | 18.8/II | 23.3/III | 31.6/III | 35.7/IV | 49.5/IV |
| | | | 191 | 191 | | |
| 186 | 185 | 184 | 184 | 184 | 179 | 176 |
| | | | 74 | 77 | 78.5 | 77 |
| 57 | 61 | 60 | 60 | 69 | | |
| | | | 50 | 49 | | |
| 40 | 38 | | | | | |
| 17.5 | 22 | 24 | | | | |
| 15 | 17 | 18 | | | | |

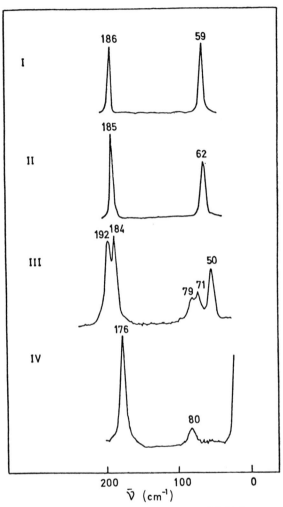

*Fig. 19.* Raman spectra of polymorphs of HgBr$_2$ at 150 K. Pressure is 0 for I, 15 for II, 31 for III, and 50 kbar for IV. [From Adams and Appleby (1977a).]

Mercuric iodide was the first material used in Raman high-pressure studies with the DAC because it is an excellent Raman scatterer. These measurements were made almost simultaneously by Postmus *et al.* (1968) and Brasch and Lippincott (1968). Adams and Appleby (1978) and Adams *et al.* (1981) have reexamined the system

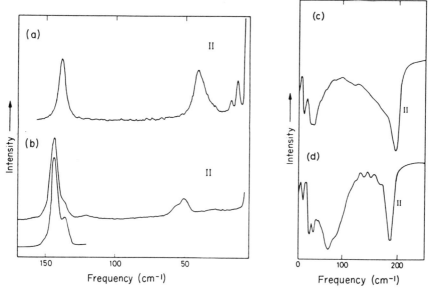

*Fig. 20.* Raman spectra of (a) $HgI_2$ (ht) at 403 K and ambient pressure (spectral slit width 0.6 cm$^{-1}$, 200 mW, 647.1 nm radiation at the sample) and (b) $HgI_2$ (II) (hp) at 295 K and 23.0 kbar (spectral slit width 1 cm$^{-1}$, 150 mW, 647.1 nm radiation at the sample); (c) IR spectra of $HgI_2$ (II) (ht) at 403 K and ambient pressure and (d) $HgI_2$ (II) (hp) at 31 kbar and ambient temperature. [From Adams *et al.* (1981).]

comprehensively and made both IR and Raman measurements at high pressures. The room-temperature red phase of $HgI_2$ converts at 127°C to a yellow phase. A yellow phase is also formed at 13 kbar when the red phase is subjected to pressure. Raman and infrared spectra show that these two yellow phases are different. Figure 20 shows the Raman and FIR spectra of the two yellow phases. Table XIII summarizes both IR and Raman results at various pressures for phase III and the high-pressure yellow phase of $HgI_2$. Beyond 75 kbar a new phase appears, which has a structural relationship to a $CdI_2$ layer type with Hg in 6-coordination, and is consistent with consequences of increased pressure.

### b. $CdX_2$ and PbFX

Adams and Tan (1981) have looked at the pressure dependencies of the Raman-active phonon modes in $CdX_2$, where X represents Cl,

*Table XIII*

Vibrational Wave Numbers ($Cm^{-1}$) for $HgI_2$ (Phases II and III) at Ambient Temperature and Various Pressures

| | Raman | | | | IR | | | | |
|---|---|---|---|---|---|---|---|---|---|
| | Phase III | | Phase II (hp) | | Phase III | | | Phase II (hp) | |
| Pressure (kbar) | 0.001 | 8.0 | 13.5 | 21.0[a] | 0.001 | 8.0 | 13.0 | 31.0 | 45.0 |
| | 17.5 | 20 | | | 25 | 25 | 41 | 40 | 40 |
| | 29 | 34 | | | | | | 48 | 48 |
| | | | 47 | 47.5 | | | | | |
| | | | ~51(sh) | ~54(sh) | | | | 56 | 56 |
| | 113.5 | 119 | ~135(sh) | ~135(sh) | | | | 76 | 82.5 |
| | | | 140.5 | 141.5 | 105(br) | 109(br) | | 93(sh) | 102(sh) |
| | | | | | | | | 186 | 182.5 |

[a] Raman experiments at significantly higher pressures were precluded by darkening of the sample; see text.

Br, and I, and in PbFX, where X represents Cl and Br. These molecules possess layer structures. Rigid layer shear modes were observed for $CdBr_2$ and $CdI_2$ at 16.1 and 15.8 $cm^{-1}$, respectively. Figure 21 shows this mode in $CdBr_2$. These modes have a higher Grüneisen constant than the intralayer modes. Table XIV summarizes results. Table XV tabulates similar results for PbFCl and PbFBr.

## 9. MISCELLANEOUS SYSTEMS

### a. Ammonium Halides

Wong and Whalley (1972) have made a Raman study of $NH_4F$ and Ebisuzaki and Nicol (1969) have examined $NH_4Cl$ under pressure by using Raman spectroscopy. At 10 kbar a disordered–ordered phase transition takes place in $NH_4Cl$. The lattice mode involving movement of the $NH_4^+$ and $Cl^-$ sublattices shows a pressure dependency of 2.65 $cm^{-1}$ $kbar^{-1}$ in the disordered (low-pressure) phase. The librational mode behaves similarly. With the exception of the $\nu_2$ bending vibration in $NH_4^+$, the internal modes all show

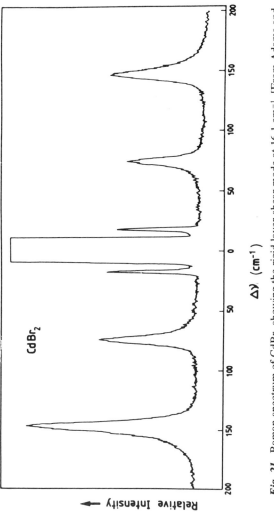

*Fig. 21.* Raman spectrum of CdBr$_2$ showing the rigid layer shear mode at 16.1 cm$^{-1}$. [From Adams and Tan (1981). Reprinted with permission from *Journal of Physics and Chemistry of Solids*, Vol. 42. Copyright 1981, Pergamon Press, Ltd.]

### Table XIV

*Raman Frequencies ($Cm^{-1}$) for Cadmium Dihalides at Various Pressures and Their Mode Grüneisen Constants[a]*

| | \multicolumn{6}{c}{$CdCl_2$} |
| | Pressure (kbar) | | | | | |
| | 0.001 | 14 | 31 | 40 | $(d\bar{\nu}_i/dp)$ | $\gamma_i$ |
|---|---|---|---|---|---|---|
| Mode $A_{1g}$ | 233.5 | 248.5 | 262.5 | 269.0 | 0.77 | 0.9 |
| Mode $E_g$ | 131.0 | 139.5 | 150.5 | 157.0 | 0.66 | 1.3 |

| | \multicolumn{6}{c}{$CdBr_2$ (4H-polytype)} |
| | Pressure (kbar) | | | | | |
| | 0.001 | 12 | 27 | 39 | $(d\bar{\nu}_i/dp)$ | $\gamma_i$ |
|---|---|---|---|---|---|---|
| Mode $A_1$ | 146.6 | 157.7 | 167.1 | 173.9 | 0.62 | 0.9 |
| Mode $E_1$ | 75.4 | 80.3 | 82.5 | 84.8 | 0.20 | 0.6 |
| Mode $E_2$ | 16.1 | 21.6 | 28.1 | 29.4 | 0.29 | 5.5[b] |

| | \multicolumn{6}{c}{$CdI_2$ (4H-polytype)} |
| | Pressure (kbar) | | | | | |
| | 0.001 | 12 | 27 | 38 | $(d\bar{\nu}_i/dp)$ | $\gamma_i$ |
|---|---|---|---|---|---|---|
| Mode $A_1$ | 111.5 | 118.9 | 128.5 | 132.9 | 0.55 | 0.85 |
| Mode $E_1$ | 43.5 | 49.9 | 54.3 | 56.4 | 0.25 | 1.0 |
| Mode $E_2$ | 15.8 | 21.0 | 26.1 | 28.9 | 0.34 | 6.00[c] |

[a] Taken in part from Adams and Tan (1981).
[b] Linear compressibility $\beta_\parallel = 3.25 \times 10^{-3}$ $kbar^{-1}$ are used.
[c] Linear compressibility $\beta_\parallel = 3.6 \times 10^{-3}$ $kbar^{-1}$ are used.

negative pressure dependencies varying from $-0.1$ to $-1.2$ $cm^{-1}$ $kbar^{-1}$ and are attributed to the increased hydrogen bonding occurring in the ordered high-pressure phases.

Ammonium iodide was studied at 130 K and pressure by Hochheimer et al. (1976). A phase change was detected by Raman spectroscopy.

## Table XV

### Raman Frequencies ($Cm^{-1}$) for PbFCl and PbFBr at Various Pressures and Their Mode Grüneisen Constants

**PbFCl**

| | Pressure (kbar) | | | | | |
|---|---|---|---|---|---|---|
| | 0.001 | 16 | 24 | 39 | $(d\bar{\nu}_i/dp)^a$ | $\gamma_i{}^c$ |
| Mode $E_g$ (1) | 240.5 | 246 | 250 | 257 | 0.44 | 0.5 |
| Mode $B_{1g}$ | 225.5 | 228 | 232.5 | 237.5 | 0.36 | 0.4 |
| Mode $A_{1g}$ | 164 | 171 | 174 | 180 | 0.42 | 0.7 |
| Mode $E_g$ (2) | 133 | 139 | 142.5 | 148.5 | 0.40 | 0.8 |
| Mode $A_{1g}$ | 106.5 | 109 | 110.5 | 114.5 | 0.205 | 0.5 |
| Mode $E_g$ (3) | 43.5 | 45.5 | 47 | 49 | 0.15 | 0.9 |

**PbFBr[b]**

| | Pressure (kbar) | | | | | |
|---|---|---|---|---|---|---|
| | 0.001 | 16 | 28 | 34 | $(d\bar{\nu}_i/dp)$ | $\gamma_i{}^d$ |
| Mode $A_{1g}$ | 116 | 123 | 127 | 129 | 0.38 | 0.7 |
| Mode $E_g$ (2) | 93 | 100 | 105 | 109 | 0.46 | 1.0 |
| Mode $A_{1g}$ | 89 | 92 | 95 | 103 | 0.36 | 0.9 |
| Mode $E_g$ (3) | 39 | 45 | 48 | 50 | 0.31 | 1.7 |

[a] Slopes of least-squares lines.
[b] $E_g$ (1) and $B_{1g}$ modes too weak to measure reliably.
[c] $X_T = 3.8 \times 10^{-3}$ kbar$^{-1}$ assumed.
[d] $X_T = 4.6 \times 10^{-3}$ kbar$^{-1}$ assumed.

## b. Ammonium Sulfate and Other Ammonium Salts

Ammonium sulfate was studied by Iqbal and Cristoe (1976). Band width dependencies on temperature and pressure were measured by using Raman spectroscopy. The authors believed that a phase-change transition to a ferroelectric phase at 223 K did not involve a soft mode.

Hamann and Linton (1975b) studied several solid ammonium salts under pressures of up to 50 kbar with the DAC and by using vibrational spectroscopy to identify changes. Among these were

***Table XVI***

*Structural Data for the NaCN and KCN Polymorphs[a]*

| Phase | Structure | Z' |
|---|---|---|
| KCN (I) | Cubic, $F_{m3m}$ ($O_h^5$) | 4 |
| KCN (III) | Cubic, $P_{m3m}$ ($O_h^1$) | 1 |
| KCN (IV) | Monoclinic, $Cm$ ($C_s^3$) | 2 |
| KCN (V) | Orthorhombic, $I_{mmm}$ ($D_{2h}^{25}$) | 2 |
| KCN (VI) | Orthorhombic, $P_{mmn}$ ($D_{2h}^{13}$) | 2 |
| NaCN (I) | Isostructural with KCN I | 4 |
| NaCN (II) | Isostructural with KCN V | 2 |
| NaCN (III) | Isostructural with KCN VI | 2 |

[a] Taken in part from Adams and Sharma (1978b).

$NH_4F$, $NH_4Cl$, $NH_4Br$, and $NH_4I$. They confirmed a previously reported phase transition in $NH_4F$. Previously identified phase transitions occurring with pressure were confirmed for $NH_4HSO_4$ and $NH_4SCN$.

### c. Alkali Metal Cyanides

The various polymorphic phases of KCN and NaCN have been investigated by using vibrational spectroscopy at high pressures (Adams and Sharma, 1978b). Table XVI lists structural data for the polymorphs of KCN and NaCN. Raman frequencies are listed in Table XVII for KCN polymorphs at various pressures. The behav-

***Table XVII***

*Raman Frequencies of $\nu_{CN}$ ($Cm^{-1}$) for KCN Polymorphs[a]*

| | Pressure (kbar) | | | | | | | |
|---|---|---|---|---|---|---|---|---|
| | 0.001 | 6.58 | 16.5 | 20.1 | 27.6 | 0.001 | 0.001 | 21.7 |
| Temperature (K) | 293 | | | | | 87 | 60 | 383 |
| Phase | I | I | I | I and IV | IV | V | VI | III |
| $\nu_{CN}$ | 2078.2[b] | 2083.5 | 2090 | 2090 2088(sh) | 2090 | 2080.0[b] | 2081.1[b] | 2090 |

[a] Based on Adams and Sharma (1978b).
[b] ±0.2 cm$^{-1}$; all others ±0.05 cm$^{-1}$.

## 9. Miscellaneous Systems

**Table XVIII**

*Pressure Dependencies of $\nu_{CN}$ Vibration in Various Polymorphs of KCN and NaCN[a]*

| $\nu_{CN}$ | $d\nu/dp$ (cm$^{-1}$ kbar$^{-1}$) | Temperature (°C) |
|---|---|---|
| KCN I   | 0.50  | 20  |
| KCN I   | 0.574 | 110 |
| KCN III | 0.45  | 110 |
| KCN IV  | 0.32  | 20  |
| NaCN II | 0.57  | —   |

[a] Taken in part from Adams and Sharma (1978b).

ior of the cyanide vibration with pressure in KCN and NaCN is shown in Table XVIII. Sodium cyanide has also been measured by Dultz *et al.* (1977) at temperatures of down to 100 K and pressures up to 6 kbar by using Raman spectroscopy as the probe. A phase transition from a disordered phase ($D_{2h}^{25}$) to an ordered phase ($D_{2h}^{13}$) was observed at a temperature of 180 K. The unit cell doubled in the lower-temperature phase, and new lattice modes appeared in the Raman spectra. The phase diagram for NaCN appears in Fig. 22, and Fig. 23 shows the spectra of the various phases. The pressure dependency of the CN$^-$ librational vibration was found to be greater than the dependency of other lattice modes.

The Raman scattering of KAg(CN)$_2$ has recently been measured to 18 kbar by Wong (1979). Two high-pressure polymorphs were identified. A dramatic change in the pressure dependency of the CN stretching vibration was noted. The pressure of other internal and external modes was also determined.

### d. Azides

Sodium azide (NaN$_3$) and thallium azide (TlN$_3$) have been examined with pressure. Sodium azide undergoes a second-order phase transition with pressure. Iqbal (1973) and Simonis and Hathaway (1974) measured the Raman spectrum of this transition. Thallium azide goes through a first-order transition as a function of pressure (Cristoe and Iqbal, 1974).

Khilji *et al.* (1982) studied KN$_3$ at pressures of up to 20 kbar by using Raman spectroscopy. No phase change was identified up to this pressure.

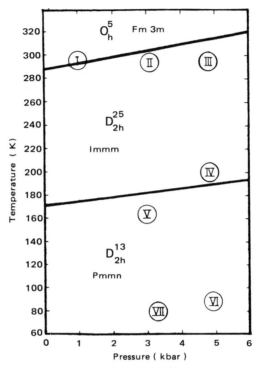

*Fig. 22.* The phase diagram of NaCN. The numbers mark pressure and temperature at which the spectra of Fig. 23 were taken. [From Dultz *et al.* (1977).]

### e. SbSI

The material SbSI is paraelectric at room temperature and possesses a $D_{2h}^{16}$ space group with four molecules per unit cell. Below 288 K it becomes ferroelectric and the space group changes to $C_{2v}^{9}$ with four molecules per unit cell. The selection rules are given as

for $D_{2h}^{16}$

$$\Gamma = 3A_g + 6B_{1g} + 6B_{2g} + 3B_{3g} + 3A_u + 5B_{1u} + 5B_{2u} + 2B_{3u}$$
$$\phantom{\Gamma =\ }(R) \quad\ (R) \quad\ \ (R) \quad\ \ (R) \quad\ (IA) \quad (IR) \quad\ \ (IR) \quad\ \ (IR)$$
$$(B_{1u} + B_{2u} + B_{3u} = \text{acoustic modes})$$

for $C_{2v}^{9}$

$$\Gamma = 8A + 7B \qquad (A + 2B = \text{acoustic modes})$$
$$\phantom{\Gamma =\ }(IR, R) \ \ (IR, R)$$

## 9. Miscellaneous Systems

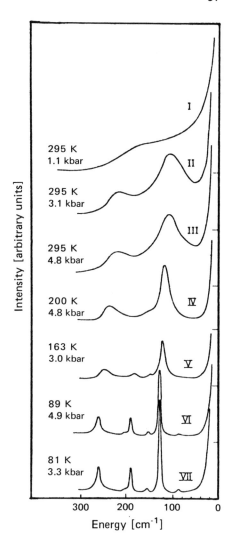

*Fig.* 23. Raman spectra of NaCN in different phases. [From Dultz *et al.* (1977).]

Peercy (1976) has examined the soft mode near 50 cm$^{-1}$ at 119.4 K by using Raman spectroscopy. Figure 24 shows the effect of pressure on the soft mode of SbSI.

### f. Nitrites (NaNO$_2$ and KNO$_2$)

Infrared spectra of KNO$_2$ and NaNO$_2$ and Raman spectra of the polymorphs of KNO$_2$ have been studied at high pressures (Adams

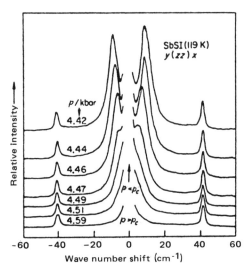

*Fig. 24.* The Raman spectra of SbSI at various pressures near $p_c$ at 119.4 K. [From Peercy (1976).]

and Sharma, 1975, 1977, 1981). Sodium nitrite ($C_{2v}^{20}$, $Z' = 1$) undergoes a phase transition at 39°C and 10 kbar, and $KNO_2$ undergoes a transition at 6.3 kbar. In the IR studies it was determined that the symmetric modes lose intensity with pressure and all bands undergo blue shifts. The Raman studies have led to conclusions concerning the nature of the polymorphs and the order or disorder in these phases.

### g. Nitrates and Carbonates

Infrared spectra of $KNO_3$ have been obtained at pressures of up to 40 kbar (Adams and Sharma, 1976a). The pressure range of $KNO_3$(III), with symmetry $R3m$ ($C_{3v}^5$) with $Z' = 1$, is very narrow and at 4 kbar and 38°C converts to $KNO_3$(IV). For this phase a new band was observed at 717 cm$^{-1}$. The 825-cm$^{-1}$ absorption shifts to 831 cm$^{-1}$ and increases in intensity with pressure. Here $\nu_3$ and $\nu_1$ show a small shift to higher frequency and a light increase in intensity. The various combinations involving the $NO_3^-$ vibration lose intensity with pressure. It was observed that the $\nu_1$ is more sensitive to pressure than $\nu_4$ and that $\nu_2$ is more insensitive. The results are consistent with a unit cell of high occupancy and low symmetry.

## 9. Miscellaneous Systems

Silver nitrate has also been investigated at high pressure (Adams and Sharma, 1976b). Adams and Sharma (1976b) have studied phases I–III and metastable phase V. In the latter phase, doublets occur in $\nu_4$, $\nu_1$, and ($\nu_1 + \nu_4$) regions and, with the shape of the $\nu_3$ envelope, are all indicative of a low site symmetry and relatively high unit cell occupancy.

The first solid carbonate study in the DAC was calcite ($CaCO_3$) (Cifruluk, 1970). The results show that $\nu_1$, normally IR inactive, appeared in the spectrum with pressure, $\nu_2$ shifted from 882 to 865 cm$^{-1}$, and splitting of $\nu_4$ occurred. Other studies were made with calcite up to 61 kbar. The authors pointed out that the spectra at higher pressures resembled that of a calcium carbonate polymorph vaterite, with a hexagonal unit cell containing two or more molecules per unit cell. Fong and Nicol (1971) have studied $CaCO_3$ up to 40 kbar and have interpreted their data in terms of two phases of calcite, II and III, occurring at 14 and 18 kbar. Raman spectra have verified that the high-pressure phases are not aragonite (Nicol and Ellenson, 1972). Only superficial studies have been made with aragonite and $MgCO_3$ under pressure (Weir *et al.*, 1959).

### h. Spinel $M'_2MO_4$ and $M''MO_4$ Compounds

The compounds $M'_2MO_4$, where $M'$ represents Na and M represents Mo and W, have a spinel structure and are cubic with a space group $O_h^7$ ($Fd3M$) and $Z' = 2$. The selection rules predict the activity

$$\Gamma_{active} = A_{1g} + E_g + 3F_{2g} + 4F_{1u}$$
$$\quad\;\;(R)\quad\;(R)\quad\;\;(R)\quad\;\;\;(IR)$$

High-pressure Raman studies have been conducted on these materials by Breitinger *et al.* (1980a,b). Table XIX summarizes the results for the Raman-active modes. It can be noted that the $d\nu/dp$ dependency for a translational mode in $Na_2MoO_4$ is intermediate between the pressure dependencies for the stretching and bending modes.

The scheelite compound $M''MO_4$, where $M''$ represents Ca, Sr, Ba, and Pb and M represents Mo and W, have a tetragonal space group $C_{4h}^6$ ($I4_1a$) and $Z' = 2$. The selection rules show the activities

$$\Gamma_{active} = 3A_g + 5B_g + 5E_g + 4A_u + 4E_u$$
$$\quad\;\;(R)\quad\;(R)\quad\;(R)\quad\;(IR)\quad\;(IR)$$

Table XX summarizes the pressure effects in the Raman spectra of $BaMoO_4$ and $BaWO_4$. The results demonstrate that the stretching

### Table XIX
#### Pressure Dependencies for $Na_2MoO_4$ and $Na_2WO_4$

| $Na_2MoO_4$ | Cm$^{-1}$ | $d\nu/dp$ (cm$^{-1}$ kbar$^{-1}$) | $Na_2WO_4$ | Cm$^{-1}$ | $d\nu/dp$ (cm$^{-1}$ kbar$^{-1}$) |
|---|---|---|---|---|---|
| $\nu_1$ ($A_{1g}$) | 892 | 0.52 | $\nu_1$ ($A_{1g}$) | 928 | 0.46 |
| $\nu_3$ ($F_{2g}$) | 809 | 0.55 | $\nu_3$ ($F_{2g}$) | 812 | 0.58 |
| $\nu_4$ ($F_{2g}$) | 381 | 0.22 | $\nu_4$ ($F_{2g}$) | 377 | — |
| $\nu_2$ ($Eg$) | 304 | 0.23 | $\nu_2$ ($Eg$) | 311 | 0.20 |
| $T'$ ($F_{2g}$)[a] | 118 | 0.37 | $T'$ ($F_{2g}$)[a] | 93 | 0.19 |

[a] $T'$ means translational mode.

vibration sensitivities are lower than in the spinel compounds, $\nu_1$ (tungstates) < $\nu_1$ (molybdates), rotary modes are sensitive for both, and translations show little pressure dependency.

Some rhenates $M'ReO_4$, where $M'$ represents $NH_4$, Na, K, Rb, and Ag, have been studied at high pressures by using Raman spectroscopy as the probe. The stretching vibrations exhibit low-pressure dependencies, and the lattice modes (especially the librations) are pressure sensitive (Breitinger et al., 1980c). Klug et al. (1982) studied the pressure dependency of three translational modes and $\nu_1$, $\nu_4$ internal modes of the $ReO_4^-$ ion by Raman spectroscopy.

### Table XX
#### Pressure Dependencies for $BaMoO_4$ and $BaWO_4$

| $BaMoO_4$ | Cm$^{-1}$ | $d\nu/dp$ (cm$^{-1}$ kbar$^{-1}$) | $BaWO_4$ | Cm$^{-1}$ | $d\nu/dp$ (cm$^{-1}$ kbar$^{-1}$) |
|---|---|---|---|---|---|
| $\nu_{1(Ag)}$ | 891 | 0.35 | $\nu_{1(Ag)}$ | 925 | 0.29 |
| $\nu_{3(Bg)}$ | 838 | 0.29 | $\nu_{3(Bg)}$ | 831 | 0.24 |
| $\nu_{3(Eg)}$ | 792 | 0.34 | $\nu_{3(Eg)}$ | 795 | 0.38 |
| $\nu_{4(Eg)}$ | 360 | 0.22 | $\nu_{4(Eg)}$ | 354 | — |
| $\nu_{4(Bg)}$ | 346 | 0.08 | $\nu_{4(Bg)}$ | 346 | — |
| $\nu_{2(Bg)}$ | 325 | 0.25 | $\nu_{2(Bg)}$ | 322 | 0.28 |
| $R'(Fg)$[a] | 189 | — | $R'(Fg)$[a] | 193 | — |
| $R'(Ag)$[a] | 140 | 0.35 | $R'(Ag)$[a] | 153 | — |
| $T'(Bg)$[a] | 136 | — | $T'(Bg)$[a] | 133 | — |
| $T'(Eg)$[a] | 108 | 0.00 | $T'(Eg)$[a] | 102 | — |
| $T'(Eg)$[a] | 76 | 0.00 | $T'(Eg)$[a] | 75 | — |
| $T'(Bg)$[a] | 79 | 0.00 | $T'(Bg)$[a] | 63 | — |

[a] $T'$ means translational mode; $R'$ is a rotational mode.

## 10. Miscellaneous Inorganic Compounds

Other scheelite-type compounds such as $TlTlX_4$, where X represents Cl and Br, have been studied with pressure. These solids all undergo low-pressure phase transitions. In $NH_4TlCl_4$ the $\nu_1$ (TlCl) vibration at 300 cm$^{-1}$ demonstrates a $d\nu/dp = 0.27$ cm$^{-1}$ kbar$^{-1}$.

Earlier Raman studies under pressure for $CaMoO_4$ and $CaWO_4$ were made by Nicol and Durana (1971). A new high-pressure phase was reported for each compound. Pressure dependencies for internal modes varied from $d\nu/dp = 0.0$ to $1.0$ cm$^{-1}$ kbar$^{-1}$.

### i. $K_2HPO_4$

The paraelectric crystals of $KH_2PO_4$ and $RbH_2PO_4$ have a $D_{2d}^{12}$ space group and with a decrease in temperature transform to a ferroelectric phase of lower symmetry $C_{2v}^{19}$. Infrared studies at pressures to 60 kbar have been conducted by Blinc et al. (1969). For both compounds the protons were found to be dynamically disordered between two possible O.H . . . O sites connecting the $PO_4^{3-}$ groups. The new phase was found at 10 kbar in which the hydrogens become ordered while the $PO_4^{3-}$ tetrahedra become disordered. In pressure studies by Peercy (1973) of $KH_2PO_4$ up to 9.3 kbar, a soft mode became overdamped with pressure application. Hamann and Linton (1975a) examined $KH_2PO_4$ to 40-kbar pressure and found that pressure had very little effect on the OH stretching vibration, and they did not find the phase transition alluded to by Blinc et al. (1969). Studies of $RbH_2PO_4$ were made at 21 kbar by using Raman spectra as the optical probe (Peercy and Samara, 1973). These results indicated a large decrease in transition temperature with pressure and a disappearance of the ferroelectric state at all temperatures for $p \geq 15.2$ kbar and a decrease with pressure of the Curie constant and dielectric constant in the paraelectric phase.

## 10. MISCELLANEOUS INORGANIC COMPOUNDS

### Oxides ($MO_2$)

*(1) $TeO_2$*  Paratellurite exists in a tetragonal $D_4^4$ space group ($P4_12_12$) and $Z' = 4$. At 9 kbar it transforms to an orthorhombic phase $D_2^4$ ($P2_12_12_1$) and $Z' = 4$. The selection rules are

$$\Gamma_{active} = \underset{(R)}{4A_1} + \underset{(IR)}{5A_2} + \underset{(R)}{5B_1} + \underset{(R)}{4B_2} + \underset{(IR, R)}{9E}$$

for $D_4^4$ and

$$\Gamma_{\text{active}} = 9A + 9B_1 + 9B_2 + 9B_3$$
$$\phantom{\Gamma_{\text{active}} = } (R) \quad (R, IR) \quad (R, IR) \quad (R, IR)$$

for $D_2^4$. Another phase transition reported at 30 kbar could not be confirmed. Raman and FIR studies at pressures of up to 45 kbar were made by Adams and Sharma (1972a). At the phase transition at 9 kbar, an $E$ mode at 122 cm$^{-1}$ splits into two components in the infrared and Raman spectra. These results aided in making proper assignments for the phonon modes in TeO$_2$.

(2) TiO$_2$ Samara and Peercy (1973) obtained Raman spectra for TiO$_2$ (rutile) at 4-kbar pressure. Rutile has a $D_{4h}^{14}$ structure with $Z' = 2$, and selection rules predict

$$\Gamma_{\text{active}} = A_{1g} + B_{1g} + B_{2g} + E_g + A_{2u} + 3E_u$$
$$\phantom{\Gamma_{\text{active}} = } (R) \quad (R) \quad (R) \quad (R) \quad (IR) \quad (IR)$$

Grüneisen parameters for each lattice mode were determined. The $B_{1g}$ phonon mode was found to have a negative pressure dependency. Results agree with those of Nicol and Fong (1971), who found the $dv/dp$ for this mode to be $-0.3 \pm 0.1$ cm$^{-1}$ kbar$^{-1}$ to 40 kbar. The mode softens slightly with decreasing temperature and may have implications relative to pressure-induced phase transitions in rutile.

(3) SiO$_2$ Several crystalline polymorphs of SiO$_2$ exist. Table XXI shows the factor group analysis for these polymorphs. Raman spectra of single crystals of $\alpha$-quartz at pressures of up to 40 kbar have been reported by Asell and Nicol (1968). The frequencies located at 128, 207, 265, 464, 697, 795, and 807 cm$^{-1}$ shift blue, with a $dv/dp$ of 0.6, 1.8, 0.5, 0.9, 0.8, 0.8, and 0.8 cm$^{-1}$ kbar$^{-1}$, respectively. Other frequencies were found to be insensitive to pressure or were not studied. Figure 25 shows a plot of Raman frequencies of $\alpha$-quartz versus pressure. Dean et al. (1982) have studied $\alpha$-quartz by Raman spectroscopy from 77–300 K and 10 kbar.

The effect of uniaxial stress on the 132-cm$^{-1}$ Raman band of $\alpha$-quartz was measured by Tekippe et al. (1973).

Infrared studies of $\alpha$-quartz, fused silica, Pyrex, and Vycor to pressures of 60 kbar were made by Ferraro et al. (1972). Pressure dependencies are tabulated in Table XXII. The intertetrahedral Si stretching at ~800 cm$^{-1}$ was found to be more sensitive to pressure

## Table XXI

### Factor Group Analysis for SiO$_2$ Crystalline Polymorphs

| SiO$_2$ polymorphs | Crystal system | Space group and number Hermann–Mauguin | Space group and number Schoen-flies | $Z_1$ | Total modes $(3nZ')$ | Activity$^a$ IR | R | IA | AC | C | Species of vibrations and number$^b$ |
|---|---|---|---|---|---|---|---|---|---|---|---|
| $a$-quartz | Hexagonal | $P3_121$ (#152) | $D_3^4$ | 3 | 27 | 12 | 12 | 0 | 2 | 8 | $4A_1 + 5A_2 + 9E$ <br> (R)   (IR)   (IR, R) |
| Stishkovite | Tetragonal | $P4_2/mnm$ (#136) | $D_{4h}^{14}$ | 2 | 18 | 4 | 4 | 3 | 2 | 0 | $A_{1g} + B_{1g} + B_{2g} + E_g + A_{2g}$ <br> $+ 2A_{2u} + 4E_u + \underline{2B_{1u}}$ |
| $\beta$-quartz | Hexagonal | $P6_222$ (#180) | $D_6^4$ | 3 | 27 | 6 | 9 | 5 | 2 | 4 | $A_1 + 3A_2 + 2B_2 + \underline{3B_1}$ <br> (R)   (IR) <br> $+ 5E_1 + 4E_2$ <br>   (IR, R)   (R) |
| $\alpha$-cristobalite | Tetragonal | $P4_12_12$ (#92) | $D_4^4$ | 4 | 36 | 12 | 21 | 0 | 2 | 8 | $4A_1 + 5A_2 + 5B_1 + 4B_2$ <br> (R)   (IR)   (R)   (R) <br> $+ 9E$ <br>   (IR, R) |
| $\beta$-cristobalite | Cubic | $Fd3m$ (#227) | $O_h^7$ | 2 | 18 | 2 | 1 | 3 | 1 | 0 | $A_{2u} + E_u + 3F_u + F_{2g} + F_{2u}$ |
| Tridymite | Hexagonal | $P6_3/mmc$ (#194) | $D_{6h}^6$ | 4 | 36 | 6 | 4 | 11 | 2 | 0 | $A_{1g} + B_{1g} + B_{2g} + E_{1g} + \underline{2E_{2g}}$ <br> $+ A_{1u} + 3A_{2u} + \underline{2B_{1u}} + 3B_{2u}$ <br> $+ \underline{5E_u} + 4E_{2u}$ |

$^a$ Activity of modes does not count degeneracy. $Z'$, see Table I for definition.
$^b$ Gerade species Raman active and ungerade species infrared active unless specified; line under species indicates inactivity.

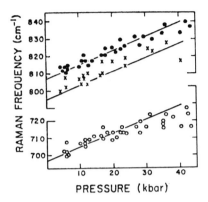

*Fig. 25.* A plot of the frequencies of the lines of the Raman spectrum of α-quartz at 695 (○), 797 (+), and 807 (●) cm$^{-1}$ versus pressure. The uncertainty of the frequency of each point is estimated to be ±5 cm$^{-1}$. [From Asell and Nicol (1968).]

than was the intratetrahedral Si–O stretching mode at ~1100 cm$^{-1}$.

The formation of stishovite ($SiO_2$) (coordination number of Si is VI) at high pressures and high temperatures in the lower mantle is discussed in Chapter 7.

(4) *MO and* $M_2O$  High-pressure studies have been made on MgO (periclase) and FeO (wüstite), and some of the results were reported in Chapter 7.

X-ray studies at high pressure on $Cu_2O$ and $Ag_2O$ have been made by Werner and Hochheimer (1982). Both materials transform from a cuprite structure to a hexagonal structure with pressure, as illustrated by the equations

$$Cu_2O \text{ (cuprite)} \xrightarrow{10^6 \text{ Pa}} Cu_2O \text{ (hexagonal)} \xrightarrow{\text{to } 1.86 \text{ Pa}} Cu_2O \text{ (}CdCl_2 \text{ structure)}$$

$$Ag_2O \text{ (cuprite)} \xrightarrow{0.46 \text{ Pa}} Ag_2O \text{ (hexagonal)}$$

(5) $MO_2$  Uranium dioxide has been studied up to pressures of 650 kbar by Benjamin *et al.* (1980–1981); it is a cubic material and at pressures between 332 and 403 kbar undergoes a phase transition to a $PbCl_2$ orthorhombic structure.

(6) $CO_2$  Xu *et al.* (1981–1982) measured infrared spectra of $CO_2$ up to 245 kbar. Up to 6 kbar the $CO_2$ spectrum was indistinguishable from gaseous $CO_2$ under ambient conditions. At 100 kbar a phase transition occurs. The pressure dependency of the combination bands between 3400 and 3900 cm$^{-1}$ was observed to shift toward higher frequency with pressure.

## 10. Miscellaneous Inorganic Compounds

**Table XXII**

*Effect of Pressure on the Principal Vibrational Frequencies of Crystalline α-Quartz, Fused Silica, Vycor, and Pyrex Glasses*[a]

| Pressure (kbar) | Frequency (cm$^{-1}$) Si–O–Si O–Si–O bending | Si stretching | Si stretching | Si–O stretching | B–O stretching |
|---|---|---|---|---|---|
| *α-quartz* | | | | | |
| 0.001 | 459 ± 2 | 698 ± 2 | 783 ± 2 | 1082 ± 3 | |
| 14.7 | 460 ± 2 | 700 ± 2 | 785 ± 2 | 1081 ± 3 | |
| 29.4 | 461 ± 2 | 702 ± 2 | 787 ± 2 | 1081 ± 3 | |
| 44.1 | 463 ± 2 | 704 ± 2 | 790 ± 3 | 1080 ± 3 | |
| 58.8 | 464 ± 2 | 707 ± 2 | 792 ± 3 | 1078 ± 3 | |
| *Fused silica* | | | | | |
| 0.001 | 475 ± 2 | | 815 ± 3 | | |
| 14.7 | 472 ± 2 | | 818 ± 3 | | |
| 29.4 | 472 ± 2 | | 820 ± 3 | | |
| 44.1 | 471 ± 2 (split at 461.5) | | 822 ± 3 | | |
| 58.8 | 471 ± 2 | | 825 ± 3 | | |
| *Vycor* | | | | | |
| 0.001 | | | 814 ± 3 | 1100 ± 3 | 1384 ± 3 |
| 14.7 | | | 817 ± 3 | 1098 ± 3 | 1386 ± 3 |
| 29.4 | | | 820 ± 3 | 1096 ± 3 | 1387 ± 3 |
| 44.1 | | | 823 ± 3 | 1095 ± 3 | 1390 ± 3 |
| 58.8 | | | 827 ± 3 | 1093 ± 3 | 1392 ± 3 |
| *Pyrex* | | | | | |
| 0.001 | | | 812 ± 5 | | 1390 ± 5 |
| 14.7 | | | 815 ± 5 | | 1392 ± 5 |
| 29.4 | | | 818 ± 5 | | 1395 ± 5 |
| 44.1 | | | 822 ± 5 | | 1400 ± 5 |
| 58.8 | | | 828 ± 5 | | 1405 ± 5 |

[a] Only vibrations showing pressure dependency are included.

(7) *$H_2O$ Ice* Figure 26 depicts the phase diagram of water ice. There are nine different phases for water ice. Bertie and Whalley (1964) measured the infrared spectra of water ice II, III, VI, and VII at appropriate pressures and temperatures. Raman spectra of water

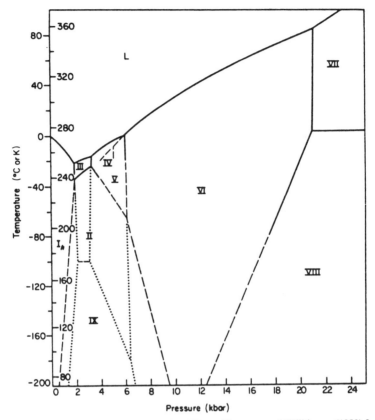

*Fig. 26.* Phase diagram of water ice. [From Sherman and Wilkinson (1982).]

ice as a function of $T$ and $p$ have been made by Wong and Whalley (1976), Marckmann and Whalley (1976), Kennedy *et al.* (1976), Sukarova *et al.* (1982), Bertie *et al.* (1968), Bertie and Whalley (1964), Bertie and Bates (1977), Whalley (1978), Sivakumer *et al.* (1978), and Block *et al.* (1965).

Water is an important material; it plays important roles in biological systems as well as in meteorology and covers a major part of the earth's surface. As a consequence, it has been extensively studied.

The spectroscopy of water ice has been used to distinguish between proton-ordered structures and proton-disordered forms. For example, ices II and IX, which are proton ordered, manifest sharp bands in the far-infrared spectra, whereas disordered ice V shows broad bands in the FIR. Similar results occur in the examination of the Raman spectra. The OH stretching modes in all the phases decrease in frequency with a pressure increase, as is expected for hydrogen-bonded systems of significant strength. Lattice modes increase in frequency with pressure. For a more detailed review of pressure and temperature studies of water ice, see Sherman (1982), Whalley (1978), Sherman and Wilkinson (1980), and Hobbs (1974).

## 11. IONIC CONDUCTORS

### a. $M_2HgI_4$ Type

The pressure-dependent Raman spectra of the fast ion conductors of the type $M_2HgI_4$, where M represents $Ag^+$, $Cu^+$, and $Tl^+$, have been measured (Grieg et al., 1977; McOmber et al., 1982). The selection rules for these materials are tabulated in Table XXIII.

Beta-$Ag_2HgI_4$ demonstrates several phases in the 0–10-kbar region. Table XXIV shows band positions in the solid at various pressures and temperatures of 25 and 60°C. At 6.1 kbar and 25°C, the HgI stretching vibration at 123 cm$^{-1}$ drops to 114 cm$^{-1}$. This may be related to an increase in coordination number from 4 to 6. A second phase transition occurs at 7.5 kbar, and at 44 kbar another phase transition takes place. Figure 27 demonstrates the Raman spectra of $\beta$-$Ag_2HgI_4$ as a function of pressure. The 24-cm$^{-1}$ mode shows a negative pressure dependency.

By means of Raman spectroscopy, $\beta$-$Cu_2HgI_4$ was followed to 24 kbar at 25° (see Fig. 28). At ~26 kbar a phase transition occurs ($\gamma$ phase), and a second new phase is found at 36 kbar ($\delta$ phase). The

*Table XXIII*

*Selection Rules for Various $M_2'HgI_4$ Salts*

| Conductor | Space group | Z' | Vibrations predicted |
|---|---|---|---|
| $\beta$-$Ag_2HgI_4$ | $S_4^2$ | 2 | $3A$ + $5B$ + $5E$ <br> (R)   (IR, R)   (IR, R) |
| $\beta$-$Cu_2HgI_4$ | $D_{2d}^{11}$ | 2 | $2A_1$ + $A_2$ + $B_1$ + $4B_2$ + $5E$ <br> (R)   (IA)   (R)   (IR)   (IR, R) |
| $Tl_2HgI_4$ | $C_2^2$ | 2 | $A + B$ |

## Table XXIV

Band Positions of Polycrystalline $Ag_2HgI_4$ at Various Pressures[a]

| Temperature (°C) | Pressure (kbar) | Position (cm$^{-1}$) | | | | | |
|---|---|---|---|---|---|---|---|
| 25 | 0 |  | 24.2 (s) | 29.5 (w) |  | 34.9 (s) | 80 (w, br) | 106 (w, br) | 122.1 (s) |
| 24 | 4 |  | 23.8 | — |  | 35.1 |  | — | 124.0 |
| 25 | 6 | 19.1 (s) |  |  | 32.8 (w) |  |  |  |
| 25 | 8 | 19.3 |  |  | 34.2 |  |  | 115.7 (m) |
| 60 | 0 | 17 (s, br) |  |  | 123 (s, br) |  | ~142 (m, sh) | 115.9 |
| 60 | 6 |  | 29 (m) | 114 (vs) |  |  |  |

[a] A gasketed diamond anvil cell with paraffin oil as a pressure transmitting liquid was used.

## 11. Ionic Conductors

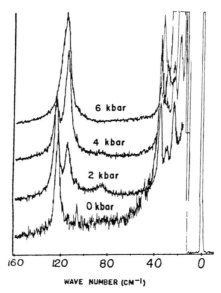

**Fig. 27.** Raman spectra of $\beta$-Ag$_2$HgI$_4$ as a function of pressure. [From Ferraro (1979).]

breathing motion of the iodide against the Cu$^+$ ions at 85 cm$^{-1}$ undergoes a more rapid blue shift than the HgI stretch at 127 cm$^{-1}$, indicative of higher anharmonicity in the CuI$_4$ stretching mode as compared to the HgI$_4$ stretch. The 37-cm$^{-1}$ vibration shows a negative pressure dependence. Figure 29 tabulates the pressure dependency of several principal frequencies in $\beta$-Cu$_2$HgI$_4$. Table XXV illustrates the $dv/dp$ values for the main bands in the three salts, including Tl$_2$HgI$_4$. Of particular interest are the 24-cm$^{-1}$ peak in Ag$_2$HgI$_4$ and the 37-cm$^{-1}$ band in Cu$_2$HgI$_4$, which show a negative pressure dependency (mode softening). Soft modes have been found in AgI, CdS, ZnO, and ZnS, all of which have a wurtzite structure (Hanson *et al.*, 1975; Mitra *et al.*, 1969; Ebisuzaki and Nicol, 1971). Hanson has suggested that for AgI the mode softening may be associated with the precursor to a phase transition that does not go to completion owing to the intervention of a different phase change.

The stability of the high-pressure phases of $\beta$-AgHgI$_4$ and $\beta$-Cu$_2$HgI$_4$ can be compared with those of AgI and CuI. The silver salts of Ag$_2$HgI$_4$ and AgI go to six-coordination at 6.2 and 4 kbar, respectively. For Cu$_2$HgI$_4$ and CuI the transitions occur at 40 and 90 kbar,

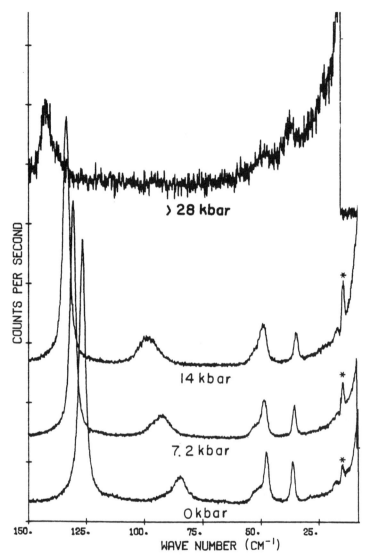

*Fig. 28.* Raman spectra of $\beta$-$Cu_2HgI_4$ as a function of pressure. [From Greig *et al.* (1977).]

## 11. Ionic Conductors

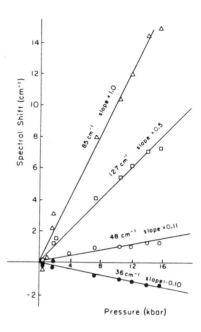

*Fig. 29.* Pressure dependency of the frequencies for the principal Raman features of $Cu_2HgI_4$. The wave number values given above each line correspond to frequencies at ambient pressure. [From McOmber et al. (1982). Reprinted with permission from *Journal of Physics and Chemistry of Solids*, Vol. 43. Copyright 1982, Pergamon Press, Ltd.]

*Table XXV*

*Pressure Dependence of Raman Peak Positions*

| Material | Peak | $\Delta\nu/\Delta p$ cm$^{-1}$ kbar$^{-1}$ |
|---|---|---|
| $Cu_2HgI_4$ | 37 | −0.10 |
| | 48 | 0.11 |
| | 85 | 1.0 |
| | 127 | 0.51 |
| $Ag_2HgI_4$ | 24 | −0.07 |
| | 29 | 0 |
| | 34 | 0.1 |
| | 81 | −0.5 |
| | 103 | Positive |
| | 122 | 0.45 |
| $Tl_2ZnI_4$ | 22 | None |
| | 29 | 0.30 |
| | 48 | ? |
| | 58 | 0.57 |
| | 80 | 1. |
| | 127 | 0.62 |

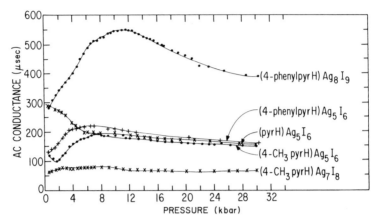

*Fig. 30.* Conductivities versus pressure for 4-phenylpyr · HI · 8AgI; 4-phenylpyr HI · 5AgI; pyr · HI; 5AgI; 4-methylpyr · HI · 5AgI; and 4-methylpyrHI · 7AgI conductors. [From Ferraro and Walling (1980).]

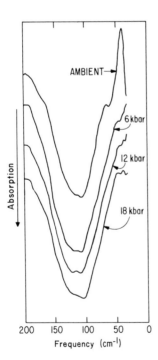

*Fig. 31.* Spectra of 4-phenylpyrHI · 8AgI versus pressure. [From Ferraro and Walling (1980).]

## 11. Ionic Conductors

respectively. Thus, the $Ag^+$ can achieve six-coordination easier than $Cu^+$, and this is consistent with the simple radius ratio considerations and with the correlation of Phillips (1970) of optical electronegativities and covalencies.

The Raman spectra for these materials have been found to be useful in screening potential ionic conductors (Hanson *et al.*, 1975; Greig *et al.*, 1976). Broad Raman bands are indicative of ionic conduction and disorder. Sharpening of the spectra is due to a more ordered structure and electronic conduction. Additionally, the results under pressure are indicators of anharmonicity occurring in the superionic conductors (Greig *et al.*, 1977). Both CuI and HgI modes are highly anharmonic, and a highly anharmonic potential may be important in providing a low-energy barrier for ion motion and ionic conductivity.

Raman shifts under pressure in $Cu_2HgI_4$ and $Tl_2HgI_4$ (McOmber *et al.*, 1982) have been used to make vibrational assignments in these compounds.

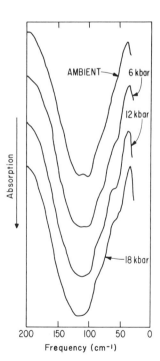

*Fig. 32.* Spectra of pyrHI · 5AgI versus pressure. [From Ferraro and Walling (1980).]

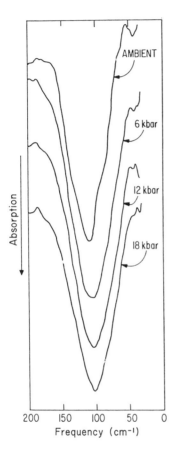

**Fig. 33.** Spectra of $\beta$-AgI as a function of pressure. [From Ferraro et al. (1980).]

### b. Organic Ammonium Silver Iodides

Pressure studies were conducted on several organic ammonium silver iodides (Ferraro and Walling, 1980). Conductivity studies under pressure indicated that several iodides demonstrated increases in conductivities with pressure. The maxima in conductivity corresponded to the formation of a doublet in the AgI vibration in the FIR (Ferraro and Walling, 1980). With further pressure the doublet began to disappear. Figure 30 shows a plot of conductivity versus pressure for several iodides, and Figs. 31 and 32 show the AgI stretching vibration of 4-phenylpyrHI · 8AgI and pyrHI · 5AgI. Silver iodide and pyrHI · 5AgI, which demonstrate no maximum in

*Table XXVI*

*Comparison of Spectra of Several AgI-Type Ionic Conductors in the AgI Phonon Mode Region with Pressure and Conductivity*

| Compound[a] | Nature of AgI phonon mode at ambient pressure | Nature of AgI phonon mode at pressure (kbar) | Conductivity results with pressure (kbar) | Remarks |
|---|---|---|---|---|
| (4$\phi$pyrHI)7AgI (I) | Main peak with shoulder at high-frequency side | Doublet (12) | Maximum (12) | Increase in disorder at outset |
| (4CH$_3$pyrHI)5AgI (I) | Main peak with shoulder at high-frequency side | Doublet with low-frequency shoulder (8) | Maximum (8) | Increase in disorder at outset |
| (4$\phi$pyrHI)·5AgI (I) | Main peak | Triplet (8) | Maximum (7) | Increase in disorder at outset |
| (pyrHI)·5AgI (I) | Doublet | Single peak (12) | Decrease | Decrease in disorder |
| (4CH$_3$pyrHI)·7AgI (I) | Doublet | Doublet (18) | No change | Little change in disorder |
| $\beta$-AgI (E) | Single peak | Single peak (18) | No change | Order maintained |
| $\beta$-AgHgI$_4$ (E) | Single peak | Single peak (12) | Slight change | Order maintained |

[a] I, ionic conductor; E, electronic conductor.

conductivity, show only a singlet with pressure. Figure 33 presents pressure data on AgI for comparative purposes. Table XXVI summarizes the pressure results.

It has been suggested that a study of these materials with pressure can be useful in studying order–disorder phenomena in ionic conductors of this type (Ferraro and Walling, 1981).

Pressure effects on other conductors are discussed in Chapter 7.

## REFERENCES

Adams, D. M., and Appleby, R. (1976). *J. Chem. Soc. Chem. Commun.*, 975.
Adams, D. M., and Appleby, R. (1977a). *J. Chem. Soc. Dalton Trans.*, 1530.
Adams, D. M., and Appleby, R. (1977b). *J. Chem. Soc. Dalton Trans.*, 1535.

Adams, D. M., and Appleby, R. (1978). *Inorg. Chim. Acta* **26**, 643.
Adams, D. M., and Sharma, S. K. (1975). *Chem. Phys. Lett.* **36**, 407.
Adams, D. M., and Sharma, S. K. (1976a). *J. Chem. Soc. Faraday Trans. 2* **72**, 1344.
Adams, D. M., and Sharma, S. K. (1976b). *J. Chem. Soc. Faraday Trans. 2* **72**, 848.
Adams, D. M., and Sharma, S. K. (1977). *Solid State Commun.* **23**, 729.
Adams, D. M., and Sharma, S. K. (1978a). *J. Phys. Chem. Solids* **39**, 515.
Adams, D. M., and Sharma, S. K. (1978b). *J. Chem. Soc. Faraday Trans. 2* **74**, 1355.
Adams, D. M., and Sharma, S. K. (1981). *J. Mol. Struct.* **71**, 121.
Adams, D. M., and Tan, T. K. (1981). *J. Phys. Chem. Solids* **42**, 559.
Adams, D. M., Appleby, R., Barlow, J., and Hooper, M. A. (1981). *J. Mol. Struct.* **74**, 221.
Asell, J. F., and Nicol, M. (1968). *J. Chem. Phys.* **49**, 5395.
Benjamin, T. M., Zou, G., Mao, H. K., and Bell, P. M. (1980–1981). *Annu. Rep. Geophys. Lab., Washington, D.C.* **80**, 280.
Berreman, D. W. (1963). *Phys. Rev.* **130**, 2193.
Bertie, J. E., and Bates, F. E. (1977). *J. Chem. Phys.* **67**, 1511.
Bertie, J. E., and Whalley, E. (1964). *J. Chem. Phys.* **40**, 1646.
Bertie, J. E., Labbé, H. J., and Whalley, E. (1968). *J. Chem. Phys.* **49**, 775.
Blinc, R., Ferraro, J. R., and Postmus, C. (1969). *J. Chem. Phys.* **51**, 732.
Block, S., Weir, C. W., and Piermarini, G. J. (1965). *Science* **148**, 947.
Bottger, C. L., and Geddes, A. L. (1967). *J. Chem. Phys.* **46**, 3000.
Brafman, O., Mitra, S. S., Crawford, R. F., Daniels, W. B., Postmus, C., and Ferraro, J. R. (1969). *Solid State Commun.* **7**, 449.
Brasch, J. W., and Lippincott, E. R. (1968). *Chem. Phys. Lett.* **2**, 99.
Breitinger, D. K., Emmert, L., and Kress, W. (1980a). *J. Mol. Struct.* **61**, 195.
Breitinger, D. K., Emmert, L., and Kress, W. (1980b). *Proc. Ontario Conf. Raman Spectrosc., 7th, Ottawa, 1980*. North-Holland Publ., Amsterdam.
Breitinger, D. K., Emmert, L., and Kress, W. (1980c). *Int. Discuss. Meet., Raman Spectrosc. Chem. Biochem., Wurtzburg, Germany, 1980*.
Buchenauer, C. J., Cerdeira, F., and Cardona, M. (1971). "Light Scattering in Solids" (M. Balkanski, ed.), p. 280. Flammerion, Paris.
Chang, I. F., and Mitra, S. S. (1968). *Phys. Rev.* **172**, 924.
Cifruluk, S. D. (1970). *Am. Mineral.* **55**, 815.
Constantino, M. S., and Daniels, W. B. (1975). *J. Chem. Phys.* **62**, 764.
Cristoe, C. W., and Iqbal, Z. (1974). *Solid State Commun.* **15**, 859.
Dean, K. J., Sherman, W. F., and Wilkinson, G. R. (1982). *Spectrochim. Acta Sect. A* **38**, 1105.
Dultz, W., Krause, H., and Winchester, L. W. (1977). *J. Chem. Phys.* **67**, 2560.
Ebisuzaki, Y., and Nicol, M. (1969). *Chem. Phys. Lett.* **3**, 480.
Ebisuzaki, Y., and Nicol, M. (1971). *J. Phys. Chem.* **33**, 763.
Ebisuzaki, Y., and Nicol, M. (1972). *J. Phys. Chem. Solids* **33**, 763.
Ferraro, J. R. (1971). "Spectroscopy in Inorganic Chemistry" (C. N. R. Rao and J. R. Ferraro, eds.), pp. 57–77. Academic Press, New York.
Ferraro, J. R. (1979). *Coord. Chem. Rev.* **29**, 1.
Ferraro, J. R., and Walling, P. (1980). *Appl. Spectrosc.* **34**, 570.
Ferraro, J. R., and Walling, P. (1981). *Appl. Spectrosc.* **35**, 217.
Ferraro, J. R., Mitra, S. S., and Postmus, C. (1966). *Inorg. Nucl. Chem. Lett.* **2**, 269.
Ferraro, J. R., Postmus, C., Mitra, S. S., and Hoskins, C. (1970a). *Appl. Opt.* **9**, 6.

# References

Ferraro, J. R., Postmus, C., Mitra, S. S., Hoskins, C., and Siwiec, E. C. (1970b). *Appl. Spectrosc.* **24**, 187.
Ferraro, J. R., Mitra, S. S., and Quattrochi, A. (1971a). *J. Appl. Phys.* **42**, 3677.
Ferraro, J. R., Horan, H., and Quattrochi, A. (1971b). *J. Chem. Phys.* **55**, 664.
Ferraro, J. R., Manghnani, M. H., and Quattrochi, A. (1972). *Phys. Chem. Glasses* **13**, 116.
Fong, M. Y., and Nicol, M. (1971). *J. Chem. Phys.* **54**, 579.
Greig, D., Joy, G. C., and Shriver, D. F. (1976). *J. Electrochem. Soc.* **123**, 588.
Greig, D., Shriver, D. F., and Ferraro, J. R. (1977). *J. Chem. Phys.* **66**, 5248.
Hamann, S. D., and Linton, M. (1975a). *Aust. J. Chem.* **28**, 2567.
Hamann, S. D., and Linton, M. (1975b). *High Temp.–High Pressures* **7**, 165.
Hamann, S. D., and Linton, M. (1976). *Aust. J. Chem.* **29**, 479.
Hanson, R. C., Fjeldly, T. A., and Hochheimer, H. D. (1975). *Phys. Status Solidi B* **70**, 567.
Hobbs, P. V. (1974). "Ice Physics." Clarendon Press, Oxford.
Hochheimer, H. D., Spanner, E., and Straach, D. (1976). *J. Chem. Phys.* **64**, 1583.
Iqbal, Z. (1973). *J. Chem. Phys.* **59**, 1769.
Iqbal, Z., and Cristoe, C. J. (1976). *Solid State Commun.* **18**, 269.
Johari, G. P., and Sivakumer, J. (1978). *J. Chem. Phys.* **69**, 5557.
Kennedy, J. M., Sherman, W. F., Treloar, N. C., and Wilkinson, G. R. (1976). *Proc. Int. Conf. Raman Spectrosc., 5th, Schultz Verlag, Freiburg, 1976.*
Khilji, M. Y., Sherman, W. F., and Wilkinson, G. R. (1982). *J. Raman Spectrosc.* **12**, 300.
Klug, D. D., Sim, P. G., and Brown, R. J. C. (1982). *J. Raman Spectrosc.* **13**, 53.
Kobashi, K., Helmy, A. A., Etters, R. D., and Spain, I. L. (1982). *Phys. Rev. B* **26**, 5996.
Lowndes, R. P., and Rastogi, A. (1976). *Phys. Rev. B* **14**, 3598.
McDevitt, N. T., Witkowski, R. E., and Fateley, W. C. (1967). *Abst. Colloq. Spectrosc. Int., Ottawa, 13th, 1967.* Adam Hilger, London.
McOmber, J. I., Shriver, D. F., Ratner, M. A., Ferraro, J. R., and Walling, P. (1982). *J. Phys. Chem. Solids* **43**, 903.
Marckmann, J. P., and Whalley, E. (1976). *J. Chem. Phys.* **64**, 2359.
Medina, F. D., and Daniels, W. B. (1976). *J. Chem. Phys.* **64**, 150.
Mitra, S. S. (1969). *Phys. Sci. Res. Pap. (U.S. Air Force Cambridge Res. Lab.)*, 69-04681.
Mitra, S. S. (1971). Air Force Cambridge Res. Lab., unpublished data.
Mitra, S. S., Postmus, C., and Ferraro, J. R. (1967). *Phys. Rev. Lett.* **18**, 455.
Mitra, S. S., Brafman, O., Daniels, W. B., and Crawford, R. K. (1969). *Phys. Rev.* **186**, 942.
Nicol, M., and Durana, J. F. (1971). *J. Chem. Phys.* **54**, 1436.
Nicol, M., and Ellenson, W. D. (1972). *J. Chem. Phys.* **56**, 677.
Nicol, M., and Fong, M. Y. (1971). *Chem. Phys.* **54**, 3167.
Nicol, M., Kersler, J. R., Ebisuzaki, Y., Ellenson, W. D., Fong, M., and Gratch, C. S. (1972). *Dev. Appl. Spectrosc.* **10**, 79.
Obriot, J., Fondére, F., and Marteau, P. (1978). *Infrared Phys.* **18**, 607.
Obriot, J., Fondére, F., Marteau, P., and Vu, H. (1979). *High Pressure Sci. Technol., AIRAPT Conf., 6th, 1977* **1**, 459.
Peercy, P. S. (1973). *Phys. Rev. Lett.* **31**, 379.

Peercy, P. S. (1976). *Proc. Int. Conf. Raman Spectrosc., 5th* (E. D. Schmid, J. Brandmüller, W. Kiefer, B. Shrader, and H. W. Schrötter, eds.). Schultz Verlag, Freiburg.
Peercy, P. S., and Samara, G. A. (1973). *Phys. Rev. B* **8**, 2033.
Phillips, J. C. (1970). *Rev. Mod. Phys.* **42**, 317.
Postmus, C., Ferraro, J. R., and Mitra, S. S. (1968). *Phys. Rev.* **174**, 983.
Postmus, C., Ferraro, J. R., Mitra, S. S., and Hoskins, C. J. (1970). *Appl. Opt.* **9**, 5.
Samara, G. A., and Peercy, P. S. (1973). *Phys. Rev. B* **7**, 1131.
Shawyer, M., and Sherman, W. F. (1978). *Infrared Phys.* **18**, 909.
Shawyer, M., and Sherman, W. F. (1982). *Infrared Phys.* **22**, 23.
Sherman, W. F. (1982). *Bull. Soc. Chim. France*, No. 9–10, pp. I347–I369.
Sherman, W. F., and Wilkinson, G. R. (1980). *Adv. Infrared Raman Spectros.* **6**, 158–336.
Sivakumer, J., Johari, G. P., and Chew, H. A. M. (1978). *Nature* **275**, 524.
Simonis, G. J., and Hathaway, C. E. (1974). *Phys. Rev. B* **10**, 4419.
Sukarova, B., Sherman, W. F., and Wilkinson, G. R. (1982). *J. Mol. Struct.* **79**, 289.
Tekippe, V. J., Ramdes, A. K., and Rodriguez, S. (1973). *Phys. Rev.* **8**, 706.
Vegard, L. (1947). *Skr. Nor. Vidensk. Akad. Kl 1: Mat. Naturvidensk. Kl.*, No. 2, 83. [Also published (1949) in *Chem. Abstr.* **43**, 4073h.]
Venugopalem, S., and Ramdas, A. K. (1973). *Phys. Rev.* **8**, 717.
Weinstein, B. A., and Piermarini, G. J. (1974). *Phys. Lett. A* **14**, 48.
Weinstein, B. A., and Piermarini, G. J. (1975). *Phys. Lett. B* **12**, 1172.
Weir, C. E., Lippincott, E. R., Valkenburg, A., and Bunting, E. N. (1959). *J. Res. Natl. Bur. Stand. Sect. A* **63**, 55.
Werner, A., and Hochheimer, H. D. (1982). *Phys. Rev. B* **25**, 5929.
Whalley, E. (1978). "High Pressure Chemistry" (H. Kelm, ed.), pp. 127–158. Reidel Publ., Dordrecht, Netherlands.
Whalley, E., Lavergne, A., and Wong, P. T. T. (1976). *Rev. Sci. Instrum.* **47**, 845.
Wong, P. T. T. (1979). *J. Chem. Phys.* **70**, 456.
Wong, P. T. T., and Whalley, E. (1972). *Rev. Sci. Instrum.* **43**, 935.
Wong, P. T. T., and Whalley, E. (1976). *J. Chem. Phys.* **64**, 2359.
Xu, J., Bell, P. M., and Mao, H. K. (1981–1982). *Annu. Rep. Geophys. Lab., Washington, D.C.* **81**, 396.

## BIBLIOGRAPHY

Bassett, W. A., and Merrill, L. (1973). The crystal structure of $CaCO_3$. II. A metastable high pressure phase of calcium carbonate. *Am. Mineral.* **58**, 1106.
Heyns, A. M., Hirsch, K. R., and Holzapfel, W. B. (1980). The effect of pressure on the Raman spectrum of $NH_4I$. *High Pressure Sci. and Technol. AIRAPT Conf., 7th* **1**, 766.
Jeanloz, R., Ahrens, T. J., Mao, H. K., and Bell, P. M. (1979). Discovery of a new phase of CaO via shock-wave and diamond anvil cell techniques. *Bull. Am. Phys. Soc.* **24**, 724.
Jeanloz, R., Ahrens, T. J., Bell, P. M., and Mao, H. K. (1979). The $B_1/B_2$ transition in calcium oxide from shock-wave and diamond cell experiments. *Annu. Rep. Geophys. Lab., Washington, D.C.* **78**, 627.

Liebenberg, D. H., Mills, R. L., Bronson, J. C., and Schmidt, L. C. (1978). High-pressure gases in diamond cells. *Phys. Lett. A* **67,** 162.

Liebenberg, D. H., Mills, R. L., and Bronson, J. C. (1980). Sound velocity and equation of state measurements in high-pressure fluid and solid helium. *High Pressure Sci. Technol. AIRAPT Conf., 7th* **2,** 609.

Pinceaux, J. P., Besson, J. M., Rimsky, A., and Weil, G. (1980). Homogenous stress conditions in the DAC: An application to GaP under high pressure. *High Pressure Sci. Technol. AIRAPT Conf., 7th* **1,** 241.

Polian, A., Besson, J. M., Chervin, J. C., and Chevy, A. (1980). Optical studies of GaS under pressure. *High Pressure Sci. Technol. AIRAPT Conf., 7th* **1,** 517.

Sugai, S., and Veda, T. (1982). High-pressure Raman spectroscopy in the layer materials 2H-$MoS_2$, 2H-$MoSe_2$ and 2H-$MoTe_2$. *Phys. Rev. B* **26,** 6554.

Wilburn, D. R., and Bassett, W. A. (1977). Isothermal compression of magnetic $Fe_3O_4$ up to 70 kbar under hydrostatic conditions. *High Temp.–High Pressures* **9,** 35.

## Ice ($H_2O$)

Abebe, M., and Walrafen, G. E. (1979). Raman studies of ice VI using a diamond anvil cell. *J. Chem. Phys.* **71,** 4167.

Yamamoto, K. (1980). Liquid figures of high-pressure ice VI. *J. Appl. Phys. Jpn.* **19,** 2291.

Yamamoto, K. (1980). Supercooling of the coexistent state of ice VII and water within ice VI region observed in diamond-anvil pressure cells. *J. Appl. Phys. Jpn.* **19,** 1841.

## Ice ($CO_2$, $NH_3$)

Hanson, R. C., Jordan, M., and Bachman, K. (1980). Studies on solid $NH_3$ and $CO_2$ to 11 GPa. *High Pressure Sci. Technol. AIRAPT Conf., 7th* **2,** 763.

## Ice (He)

Pinceaux, J. P., Maury, J. P., and Besson, J. M. (1979). Solidification of helium at room temperature under high pressure. *J. Phys. Lett. Orsay, Fr.* **40,** 307.

## Elements

Block, S., and Piermarini, G. J. (1973). The melting curve of sulfur to 300°C and 12 kbar. *High Temp.–High Pressures* **5,** 567.

Ekberg, S., Cromer, D. T., Lesar, R., and Shaner, J. (1980). Solid nitrogen at pressures to 410 kbar. *High Pressure Sci. Technol. AIRAPT, Conf., 7th* **2,** 612.

Pinceaux, J. P., Maury, J. P., and Besson, J. M. (1979). Solidification of helium at R.T. under high pressure. *J. Phys. Lett. Orsay, Fr.* **40,** 307.

# CHAPTER 5
# COORDINATION COMPOUNDS

The DAC has played a very important role in the field of coordination chemistry. The visible and, in general, the far-infrared regions have been the diagnostic tools used to record the changes. The response to pressure for coordination compounds should be similar to those for inorganic molecules. Certain vibrations will be sensitive to pressure in contrast to others. Generally, blue shifts will be produced. The responses of vibrations to pressure have been used to aid in making assignments in the FIR. New pressure-stable phases will occur. Perhaps more important are the changes relative to spin states, oxidation states, and changes in geometry occurring with pressure. Sufficient research on the solid state has now been accomplished to test the theories of Pearson and Bader, which applied primarily to structural interconversions in solution. This chapter summarizes the work done on coordination compounds under high pressure and compares results for solids with the predictions of Pearson and Bader.

## 1. SOLID-STATE STRUCTURAL CONVERSIONS

The geometric structure of a molecule depends on the magnitude of the energy barrier that prevents conversion into any of its geometric isomers.

## 1. Solid-State Structural Conversions

In an extensive and impressive group of papers, Pearson (1969a,b, 1970, 1971a,b, 1972, 1976) and Bader (1960, 1962) have presented a series of symmetry rules that can be used to predict the most stable structure of the molecule (Pearson, 1969a, 1970, 1971a), its structural rigidity, and its mode of reaction (Pearson 1969b, 1971b, 1972). The rules, which are an extension of the work of Bader (1960, 1962), are based on the use of perturbation theory and group theory to evaluate the effect of a vibrational distortion on the ground-state geometric configuration of a molecule. The theory supporting these symmetry rules has been extensively developed during the past few years (Pearson 1969a,b, 1970, 1971a,b, 1972, 1976; Longuet-Higgins, 1956). However, in general, experimental verification of the rules has been based on molecular interconversions or reactions that occur either in solution or in the isolated gaseous state. The reliability of these orbital symmetry rules in predicting a stereochemically rigid or flexible geometric structure in the solid state has not as yet been experimentally confirmed.

Pressure studies of solid coordination compounds have provided evidence that indicates that these symmetry rules are useful for predicting solid-state structural interconversions between solid geometric isomers at high pressure. Furthermore, the studies provide a new approach to the classification of the various effects that are observed in coordination compounds under high pressure.

In the approach developed by Pearson, it is first assumed that all first-order structural distortions of the molecular geometry have occurred. These distortions include any first-order Jahn–Teller distortions required to produce a nondegenerate electronic ground state and any vibrational distortions that can occur along the totally symmetric normal vibrational modes. On the basis of this assumption, the energy of an initial molecular configuration in the presence of a distortion can be expressed as

$$E = E_o + \frac{1}{2} Q^2 \int \psi_o \left| \frac{\partial^2 v}{\partial Q^2} \right| \psi_o \, d\tau + \frac{\Sigma [Q \int \psi_o | \partial v / \partial Q | \psi_k \, d\tau]^2}{E_o - E} \quad (1)$$

$$E = E_o + f_{oo} Q^2 + f_{ok} Q^2 \quad (2)$$

where $Q$ is a measure of the magnitude of the displacement of the initial molecular configuration along a normal coordinate, which is then designated as the $Q$ coordinate, and $v$ the nuclear–nuclear and nuclear–electronic potential energy. The remaining symbols have their usual meanings and are discussed later. The reader should consult Pearson (1976) and Bader (1960, 1962) for further details on the derivation and use of this expression.

In Eqs. (1) and (2), the first term $E_o$ is the minimized energy of the initial undistorted molecular configuration. The term $f_{oo}Q^2$ represents the distortion-induced change in the energy of the initial electronic configuration and is always positive because the initial energy is minimized for the undistorted initial molecular configuration. This term provides a restoring force that would tend to remove the distortion and return the molecule to its initial configuration. The term $f_{ok}Q^2$ is always negative because it, in effect, changes the initial wave function $\psi_o$ to fit the nuclear coordinates of the distorted molecular configuration. The sum of the second and third terms is the experimental force constant for the normal coordinate $Q$.

Three cases are possible, depending on the relative magnitudes of the values of $f_{oo}$ and $f_{ok}$. If $f_{oo} \gg f_{ok}$, the initially chosen molecular configuration is stable and no distortion is expected. If $f_{oo} \simeq f_{ok}$, the initial molecular configuration is unstable and may be spontaneously distorted along the normal coordinate $Q$ to a new molecular configuration. Finally, $f_{oo} < f_{ok}$, the initial molecular configuration is unstable and will spontaneously change via normal coordinate $Q$ to a relatively more stable configuration. From these results we can see that a knowledge of the magnitude of $f_{ok}$ can provide us with some information about the relative stability of the initial molecular configuration.

At this point, as discussed in detail by Pearson (1969b, 1971b, 1972), some rather extensive approximations must be made in order to evaluate the probable magnitude of $f_{ok}$. The first of these approximations involves limiting the summation in the $f_{ok}$ term to only the one or two lowest excited electronic states (represented by $\psi_k$) available to the molecular configuration. This approximation will only be good if the energy of the remaining terms is much greater than $E_o$. The second approximation involves replacing the total ground-state electronic wave function $\psi_o$ with the highest occupied molecular orbital $\psi_{HOMO}$ for the configuration. Similarly, the total first excited-state wave function $\psi_k$ is replaced by the lowest unoccupied molecular orbital $\psi_{LUMO}$ for the configuration, etc.

We can determine whether the integrals in the $f_{ok}$ terms are zero or nonzero by making use of the symmetry of the wave functions and the symmetry of the operator $\delta\nu/\delta\psi$, which will be the same as

## 1. Solid-State Structural Conversions

that of the normal coordinate $Q$. The integral will be nonzero only if the direct symmetry product contains the totally symmetric irreducible

$$\Gamma_{\psi_{\text{HOMO}}} \times \Gamma_Q \times \Gamma_{\psi_{\text{HOMO}}} \tag{3}$$

representation. In Pearson's papers, this requirement is stated in terms of the transition density $p_{ok}$, which is the product $\psi_o \psi_k$ of the two wave functions and represents the amount of electronic charge transferred within the molecule as a result of nuclear motion. In terms of the transition density, the integral will be nonzero only if the symmetry of the transition density is the same as that of $Q$, i.e.,

$$\Gamma_{p_{ok}} = \Gamma_{\psi_o} \times \Gamma_{\psi_k} = \Gamma_{\psi_{\text{HOMO}}} \times \Gamma_{\psi_{\text{LUMO}}} = \Gamma_Q \tag{4}$$

Apart from the symmetry requirement, the value of $f_{ok}$ will depend on the energy difference $E_o - E_k$. Pearson has suggested that an energy gap of the order of 4 eV between the HOMO and LUMO is small enough to indicate the possibility of a structural instability. Hence, with the proper symmetry and with a small enough energy gap, a change in molecular configuration may occur. However, there is an additional requirement for the occurrence of a molecular interconversion, namely, that distortion along the normal coordinate $Q$ must, if continued, lead to an alternative structure.

In summary, there are three basic requirements that, if satisfied, would lead to a possible structural interconversion. First, the symmetry product must be correct; i.e., the symmetry of the transition density $p_{ok}$ must be the same as that of a normal coordinate $Q$. Second, the energy gap between the HOMO and one or more of the excited-state molecular orbitals must be of the order of 4 eV or less. Finally, the normal coordinate $Q$ must lead to a viable alternative structure.

These ideas, as developed by Pearson, may be useful in predicting whether or not a structural interconversion will occur in a solid that is subjected to high pressures. It should be noted that, in general, bending force constants are much smaller than stretching force constants, and, as a result, for normal modes involving vibration, $f_{ok}$ is likely to be larger than $f_{oo}$, which, in this case, should be small. Thus, molecular arrangements involving bending modes and hence changes in bond angles are more likely to occur than those involving stretch modes.

On the basis of the preceding arguments, a solid complex is more likely to show a structural interconversion along a normal coordinate at high pressure if the symmetry of that coordinate is the same as that of the transition density $p_{ok}$ as determined for the HOMO and LUMO of the complex. The lack of such agreement may not prevent such an interconversion, but it might make an alternative structural change more likely. We must also, at this point, consider the changes in the energy gap between the HOMO and the excited-state molecular orbitals at high pressure. In general, these energies are expected to shift with the application of pressure.

### a. Structural Interconversion in the Solid State

As already emphasized, most structural interconversions have been studied in solution. Nuclear magnetic resonance spectroscopy has been an ideal tool for such studies because the equilibrium established between labile structures often can be shifted in favor of one structure by a change in temperature. Until recently, few high-pressure studies of solid-state structural interconversions of complexes have used vibrational and electronic spectroscopy. The Mössbauer effect has, however, been used extensively by Drickamer and Frank (1973).

It is reasonable to assume that solid-state interconversions involve considerably larger energy effects than those observed in solution. For solid complexes, in addition to the symmetry effects discussed earlier, molecular packing, lattice forces, ligand flexibilities, metal–ligand bond distance, d–d electronic transition energies, orbital overlap and orientation effects, and hydrogen bonding, among other factors, must also be considered. High pressure is known to affect many of these factors (Drickamer and Frank, 1973; Ferraro *et al.*, 1971; Drickamer, 1965) and will favor the structure with a smaller packing volume. High-pressure effects are observed to shorten the metal–ligand bond distance and to increase the average ligand field strength $10Dq$ (Ferraro *et al.*, 1971; Drickamer, 1965; Long and Ferraro, 1974). In cases involving high-spin complexes, this increase in ligand field strength may be sufficient to overcome the electron spin pairing energy and produce a low-spin complex. See also Bargeron *et al.* (1971).

Of particular interest is the effect of high pressure on the infrared absorption bands of a solid complex. A reduction in the metal–ligand bond distance shifts the vibrational bands to higher energy.

For bending modes, which might possibly transform one structure into another, the effects of pressure may be smaller and the associated band might conceivably shift to a lower energy. It is also possible that at high pressure, normally forbidden modes might become allowed (in a lower site symmetry), and if this mode yielded a structural interconversion, the conversion might then become allowed. Thus, it is of interest to examine the solid-state rigidity of various molecules with differing stereochemical configurations at high pressure.

### b. Four-Coordinate Complexes

Pearson (1976) has analyzed transition-metal $MX_4$ complexes and has formulated structural predictions, which are tabulated in Table I. Several four-coordinate solid complexes have been subjected to pressure. The complexes dichloro- and dibromobis(benzyldiphenylphosphine)nickel(II) can be prepared as red and green isomers [see Browning et al. (1962)]. The red complexes are diamagnetic square-planar forms of the compounds. The green bromide isomer possesses a magnetic moment of 2.70 $\mu\beta$ (bohr magneton) at room temperature, and it has been demonstrated by x-ray analysis (Kilbourn et al., 1963; Kilbourn and Powell, 1970) that the

*Table I*

*Structural Predictions of Stable Structures for $MX_4$ Complexes*

| d Orbital occupation | High spin | Low spin |
|---|---|---|
| $d^0$ | $T_d$ | $T_d$ |
| $d^1$ | $D_{2d}$ | $D_{2d}$ |
| $d^2$ | $T_d$ | $D_{2d}$ |
| $d^3$ | $D_{2d}$ | $<D_{2d}$ |
| $d^4$ | $D_{2d}$ | $D_{2d}$ |
| $d^5$ | $T_d$ | $<D_{2d}$ |
| $d^6$ | $D_{2d}$ | $D_{4h}$ |
| $d^7$ | $T_d$ | $D_{4h}$ |
| $d^8$ | $D_{2d}(T_d)$[a] | $D_{4h}$ |
| $d^9$ | $D_{2d}$ | $D_{2d}$ |
| $d^{10}$ | $T_d$ | $T_d$ |

[a] $NiCl_4^{2-}$ high spin sometimes found in $D_{2d}$ or $T_d$.

unit cell has one square-planar and two tetrahedral nickel atoms and is, therefore, distorted toward the square planar. The green chloride isomer has a moment of 3.23 $\mu\beta$, and spectroscopic properties reveal that the nickel has a tetrahedral environment (Browning et al., 1962). The green isomers were subjected to pressure in a DAC and followed in the electronic and far-infrared regions of the electromagnetic spectrum (by Ferraro et al., 1973). The results are depicted in Fig. 1. The green, tetrahedral chloride (Ni(BzPh$_2$P)$_2$Cl$_2$) isomer remains tetrahedral, while the green, distorted bromide (Ni(BzPh$_2$P)$_2$Br$_2$) isomer is transformed by pressure to the purely square-planar red isomer. A change in spin state occurs along with the geometrical structural change.

In another high-pressure study (Long and Ferraro, 1973; Long and Coffen, 1974), it was possible to irreversibly convert the paramagnetic violet pseudotetrahedral nickel complex Ni(Qnqn)Cl$_2$ into its yellow, paramagnetic binuclear (Ni(Qnqn)Cl$_2$)$_2$ isomer. In these complexes, the Qnqn ligand is *trans*-2-(2'-quinolyl)methylene-3-quinuclidinone.

Both the yellow and violet isomers have been prepared (Long and Ferraro, 1973; Long and Coffen, 1974), and the x-ray structure (Long and Schlemper, 1974) of the yellow binuclear isomer has revealed two bridging and two terminal chlorine ligands and bidentate coordination for Qnqn. The application of pressure to the violet monomeric complex causes the two nickel–chloride nonbonded distances to decrease to a point at which the two additional bridging chlorine bonds are formed, and the yellow binuclear complex results. The spectrum of the complex clearly reveals the irreversible changes in both the $\nu_{\text{Ni-Cl}}$ and $\nu_{\text{Ni-N}}$ vibrational bands as a function of pressure. The electronic absorption spectrum of the violet isomer also reveals the expected changes in the d–d bands at high pressure.

This is the first example of such an irreversible pressure-induced structural transformation discovered. The irreversibility of this transformation may result from the bond energy of the two additional chlorine bridging bonds, which would make the reverse trans-

# 1. Solid-State Structural Conversions

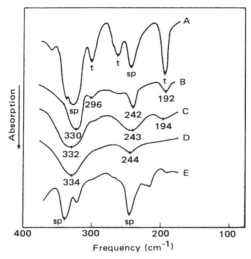

**Fig. 1.** Skeletal vibrations in the green and red isomers of solid Ni(BzPh$_2$P)$_2$Br$_2$ at ambient and high pressure; t = tetrahedral; sp = square planar. Green isomer: A, ambient pressure; B, 4 kbar; C, 12 kbar; D, 20 kbar. Red isomer: E, ambient pressure (pure sp isomer). [From Ferraro *et al.* (1973).]

formation thermodynamically unfavorable. This transformation involves both a change in coordination number and a change in coordination geometry.

The room-temperature preparation of [(CH$_3$)$_2$CHNH$_3$]$_2$CuCl$_4$ has been found from x-ray studies (Anderson and Willett, 1974) to contain one copper ion in square-planar configuration and two copper ions arranged in tetrahedrally distorted configurations. The crystal is held together by hydrogen bonding from the isopropylammonium ions. At high pressures, the coordination geometry of the two tetrahedrally distorted copper ions is reversibly converted to a square-planar geometry (Willett *et al.*, 1974). The conversion is observed as a change in the $\nu_{Cu-Cl}$ and $\nu_{ClCuCl}$ vibrational bands. Confirmation for the conversion was also found in the change occurring in the electronic region. A similar structural conversion is also found in Cs$_2$CuCl$_4$ and Cs$_2$CuBr$_4$ (Wang and Drickamer, 1973). These compounds exhibit a geometric structural change with no change in coordination number or spin state.

These studies have led to conclusions that a distorted structure for a CuCl$_4^{2-}$ salt distorted in the direction of the complex with the smaller volume is important for the structural conversion to occur (Ferraro, 1976). For these salts it was possible to convert the salts into square-planar geometry with pressure if the starting phase was distorted. It should be noted that salts with counterions are capable of strong hydrogen bonding and tend to produce geometrical structures with square-planar symmetry. It can be considered that hydrogen bonding is a form of internal pressure occurring in the solid. Table II illustrates the correlation table for the infrared-active stretching mode in $T_d$ symmetry and its lower symmetry analogs. From Table II it can be observed that the $F_2$ stretching mode is followed from $T_d \rightarrow D_{2d}$, $D_{4h} \rightarrow D_{2h}$ symmetry. In the $D_{2h}$ symmetry the single stretching mode splits into $B_{2u}$ and $B_{3u}$, with only minor frequency changes occurring. However, in going from the stretching mode ($B_2$) in $D_{2d}$ symmetry, significant shifts are observed because the lowering of symmetry to $D_{4h}$ and $D_{2h}$ converts the $B_2$ vibration to $A_{2u}$ and $B_{1u}$, respectively, and now the modes correspond to out-of-plane bends with lower energy than the stretching modes. In this way the far infrared of these complexes was understood and used for diagnostic purposes in interpreting the effects of pressure or of hydrogen bonding. The visible region was also followed. As can be seen in Table III, the electronic absorptions increase in frequency as we progress from a quasi-$T_d$ to a $D_{2h}$ geometry. Concurrently, the *trans*-Cl–Cu–Cl angle approaches 180°, and the volume decreases.

Pressure studies by Ferraro and Sherren (1978) of $MCl_4^{2-}$ ions were made. The salts had counterions capable of hydrogen bonding (considered to be an internal pressure effect) and those in which hydrogen bonding was not possible. It was found that hydrogen

*Table II*

Correlation of $T_d$ Symmetry with $D_{2d}$, $D_{4h}$, and $D_{2h}$ Symmetries for the $F_2$ Stretching Vibration[a]

| $T_d$ | $D_{2d}$ | $D_{4h}$ | $D_{2h}$ |
|---|---|---|---|
| $F_2(\nu)$ | $B_2(\nu)$ —— $E(\nu)$ | $A_{2u}(\pi)$ —— $E_u(\nu)$ | $B_{1u}(\pi)$ $B_{2u}(\nu)$ $B_{3u}(\nu)$ |

[a] $\nu$ = stretching mode; $\pi$ = out-of-plane bending.

**Table III**

*Several Parameters for $R_2CuCl_4$ Salts[a]*

| Salt | Trans Cl–CuCl angle (average°) | Density | CuCl infrared modes ($cm^{-1}$) $D_{2d}(e)$ $D_{2d}(B_2)$ | | Electronic absorptions ($cm^{-1}$) |
|---|---|---|---|---|---|
| $Cs_2CuCl_4$ | 124 | — | | | 9100 |
| $[(CH_3)_4N]_2CuCl_4$ | 128 | 1.40 | 292 | 257 | 9000 |
| $[\phi(CH_2)_2NH(CH_3)H]_2CuCl_4$ (yellow, 80°) | 131 | 1.37 | | | 9100 |
| $\phi CH_2N(CH_3)_3]_2CuCl_4$ | 132 | 1.36 | | | |
| $[(CH_3)_2CHNH_3]CuCl_4$ [$(IPA)_2CuCl_4$] (yellow, 60°) | 135 | | 290 | 232 | 10000 |
| $[(C_2H_5)_2NH_2]_2CuCl_4$ [$(DEA)_2CuCl_4$] (yellow, 43°C) | 135 | 1.33 | 295 | 220 (sh) | 10200, 7300 |
| $[(C_2H_5)_3NH]_2CuCl_4$ | 137 | 1.33 | | | |
| $[C_6H_5CH_2CH(CH_3)NH_2CH_3]_2CuCl_4$ | 138 | 1.43 | | | 10200 |
| $[C_{13}H_{19}N_2OS]_2CuCl_4$ | 143 | | | | |
| | | | $D_{2h}(B_{2u}, B_{3u})$ $D_{2h}(B_{1u})$ | | |
| $[(CH_3)_2CHNH_3]_2CuCl_4$ [$(IPA)_2CuCl_4$] (green, RT) | 155($\frac{2}{3}$) 180($\frac{1}{3}$) | 1.50 | 301,279 | 181 | 11000–13000 |
| $[(C_2H_5)_2NH_2]_2CuCl_4$ [$(DEA)_2CuCl_4$] (green, RT) | 162 | | 282,287 (sh) | 186 | 15300, 12900, 9900 |
| $[(CH_3)_2CHNH_3]_2CuCl_4$ [$(IPA)_2CuCl_4$] (yellow, HP) | ~180[b] | | 281,295 (sh) | 193 | |
| $[C_2H_5NH_3]_2CuCl_4$ | 180 | 1.70 | 278,294 (sh) | 182 | |
| $[\phi(CH_2)_2NH(CH_3)H]_2CuCl_4$ (green, LT) | 180 | 1.40(?) | | | 16900, 14300, 12500 |
| $Pt(NH_3)_4CuCl_4$ | 180[b] | — | | | 14300 |

[a] Abbreviations: RT = high temperature; LT = low temperature; HP = high pressure; sh = shoulder.
[b] Estimated.

*Table IV*

*Summary of Pressure Effects on $R_2MCl_4$ Complexes*

| Counterion[a] | | Mn(II) | Fe(II) | Co(II) | Ni(II) | Cu(II) |
|---|---|---|---|---|---|---|
| | | $d^5$ | $d^6$ | $d^7$ | $d^8$ | $d^9$ |
| $R_4N^+$ | | $T_d$ | $T_d$ | $T_d$ | $T_d$ | $D_{2d}$ |
| $R_3NH^+$ | Increase in | | | | | |
| $R_2NH_2^+$ | internal | | | | | |
| $RNH_3^+$ ↓ | pressure[b] | $O_h$ | $O_h$ | $D_{2d}$ | $O_h$[c] | $D_{2d}$ and $D_{2h}$ |
| | | | | ↓ | | ↓ |
| $R_3NH^+$ | Increase in | | | | | |
| $R_2NH_2^+$ | external | | | | | |
| $RNH_3^+$ ↓ | pressure | | | $D_{2d}$ | | $D_{2h}$ |

[a] R = alkyl group.

[b] Internal pressure is considered brought about by increased hydrogen bonding as one goes from $R_4N^+$ to $RNH_3^+$.

[c] Stoichiometry involves $NiCl_3^{-3}$ from solution; from melt, polymeric octahedral structures are obtained with the stoichiometry of $R_2NiCl_4$ (Ferraro and Sherren, 1978).

bonding stabilizes an octahedral geometry for $Mn^{2+}$ and $Fe^{2+}$. The $Ni^{2+}$ solids isolated from solution had the stoichiometry of $RNiCl_3$ and contained $NiCl_6$ octahedra, and from melts complexes of the type $R_2MCl_4$ are formed.

An examination of Table IV shows that in $MX_4$ complexes with $d^8$, $d^9$ electronic systems, on which the largest number of pressure studies have taken place, the experimental results are compatible with the predictions made by Pearson's treatment.

### c. Five-Coordinate Complexes

For $MX_5$ complexes of transition metals, Pearson has concluded that ready interconversion occurs between $C_{4v}$ and $D_{3h}$ symmetries. This is certainly verified in solution studies. The energy needed to interconvert the structures $TBP(D_{3h})$ and $C_{4v}(SqPy)$ is quite small in solution (Muetterties, 1970a). It would be expected that in the solid state more energy would be expected to interconvert these structural geometries. Several systems in the solid state have been investigated to ascertain pressure effects on the five-coordinate complexes.

## Table V

*Structural Inferences from Pressure Effects for Several Ni(II), Pd(II), Pt(II), and Co(II) Complexes Containing Ligands Varying from Tetradentate to Monodentate*[a]

| Complex[b] | Type of ligand | $d\nu/dp$ (cm$^{-1}$ kbar$^{-1}$) | Structure |
|---|---|---|---|
| [NiLX]Y (24) | Tetradentate | 33–70 | TBP |
| [PdLX]Y (2) | Tetradentate | 33–81 | TBP |
| [PtLX]Y (1) | Tetradentate | 27 | TBP |
| [CoLX]Y (1) | Tetradentate | 7 | SQP |
| [NiLX$_2$] (3) | Tridentate | 9–32 | Distorted, TBP–SQP |
| [NiL$_2$X]Y (5) | Bidentate | 9–32 | Distorted, TBP–SQP |
| [CoL$_2$X] (2) | Bidentate | Very slight shift | SQP |
| [NiL$_3$X$_2$] (6) | Monodentate | 8–29 | Distorted, TBP–SQP |
| [CoL$_3$X$_2$] (2) | Monodentate | 8–23 | Distorted, TBP–SQP |

[a] Abbreviations: TBP = trigonal bipyramid; X = halide or pseudohalide ion; SQP = square pyramid; Y = polyatomic anion.

[b] Numbers in parentheses indicate number of compounds studied.

An x-ray diffraction study by Raymond et al. (1968) of the (Cr(en)$_3$)(Ni(CN)$_5$) · 1.5 H$_2$O complex has shown that its unit cell contains two crystallographically independent (Ni(CN)$_5$)$^{3-}$ ions, one with a regular square-pyramidal geometry and one with a distorted trigonal-bipyramidal geometry. Dehydration of the complex converts all of the (Ni(CN)$_5$)$^{3-}$ ions to the square-pyramidal geometry (Raymond et al., 1968). When this compound was subjected to pressures of ~7 kbar at 78 K, the coordination geometry of the trigonal-bipyramidal (Ni(CN)$_5$)$^{3-}$ ion was converted reversibly to the square-pyramidal geometry (Basile et al., 1974). The IR spectrum of this compound at ambient and high pressure is presented in Fig. 2. In order to prevent the dehydration of the complex at high pressure (presumably a result of localized heating produced by the 6× beam condenser used with the pressure cell), these studies were made at 78 K.

An extensive high-pressure study of many five-coordinate nickel(II) complexes with ligands ranging from monodentate to tetradentate has revealed several nonrigid structures in the solid state (Ferraro et al., 1971). The results for several metal ions are presented in Table V and reveal that "tripodlike" tetradentate ligands prefer the trigonal-bipyramidal structure. The importance of the larger number

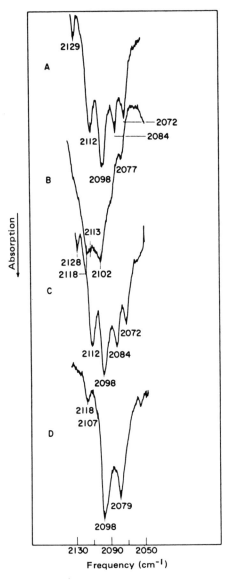

*Fig. 2.* The cyanide stretching vibrational bands in $[Cr(en)_3][Ni(CN)_5] \cdot 1.5\ H_2O$: A, at 78 K and ambient pressure; B, at 78 K and ~7 kbar; C, at 78 K and ambient pressure after release of high pressure. Spectrum D is that of $[Cr(en)_3][Ni(CN)_5]$ at ambient temperature and pressure. [Reprinted with permission from Basile *et al.* (1974). Copyright 1974 American Chemical Society.]

## 1. Solid-State Structural Conversions

*Table VI*

*Pressure Dependency of* $ML_3X_2$ *Compounds*

| Complex | Structure by x-ray data | $dv/dp$ $(cm^{-1}\ kbar^{-1})$ |
|---|---|---|
| $Ni(Me_3P)_3Br_2$ | Distorted, TBP–SqPy | 19 |
| $Ni(Me_3P)_3I_2$ | Undetermined | 35 |
| $Ni(Ph_2PH)_3Cl_2$ | Distorted, TBP–SqPy | 29 |
| $Ni(Ph_2PH)_3Br_2$ | Distorted, TBP–SqPy | 27, 29 |
| $Ni(Ph_2PH)_3I_2$ | Distorted, TBP–SqPy | 17, 27 |
| $Ni(Ph_2PMe)_3(CN)_2$ | Distorted, TBP–SqPy | 20 |
| $Co(Ph_2PH)_3Br_2$ | Distorted, TBP–SqPy | 8 |
| $Co(Ph_2PH)_3I_2$ | Distorted, TBP–SqPy | 23 |

of chelate rings and the increased entropy and free energy of formation for tetradentate ligand complexes of the type (NiLX)Y are indicated by the more numerous trigonal-bipyramidal structures. As the number of chelate rings is reduced, stability decreases, and the tendency to form intermediate five-coordinate complexes results (Ferraro and Nakamoto, 1972). $NiL_3X_2$ complexes with no chelate rings are unstable and dissociate in solution, whereas the application of high pressures tends to distort these solids toward the distorted intermediate five-coordinate geometry (Table VI).

Ferraro and Nakamoto (1972) reported on the effects of pressure on Ni(II) and Co(II) five-coordinate complexes of the type $ML_3X_2$. The ligands in these complexes ranged from tetradentate to monodentate. All of the complexes demonstrated limiting structures between TBP or square-pyramidal. As in previous studies, as the amount of chelation decreases, the tendency to form square-pyramidal structure increases. Table VII shows results.

Other predictions indicated by Pearson on the liability of the five-coordinate state are demonstrated to hold as well in the solid state.

### d. Six-Coordinate Complexes

To date, changing the coordination number or geometry of an octahedral (or close to octahedral) complex at high pressure has proven unsuccessful. However, several octahedral high-spin complexes have been reversibly converted, at least in part, to the analo-

**Table VII**

*Comparison of Pressure Dependency for Several Nickel(II) Complexes in which the Ligand Varies from a Tetradentate to a Monodentate Type*

| Complex | Type of ligand | $d\nu/dp$ (cm$^{-1}$ kbar$^{-1}$) | Structure |
|---|---|---|---|
| [NiLX]Y | Tetradentate | 33–70 | TBP |
| [NiLX$_2$] | Tridentate | 9–32 | Distorted, TBP–SqPy |
| [NiL$_2$X]Y | Bidentate | 9–32[a] | Distorted, TBP–SqPy |
| [NiL$_3$X$_2$] | Monodentate | 8–29 | Distorted, TBP–SqPy |

[a] [Ni(TEP)$_2$I]I shows a $d\nu/dp$ of 40 cm$^{-1}$ kbar$^{-1}$, but this complex may have been converted to a TBP structure with pressure. TEP = (C$_2$H$_5$)$_2$P(CH$_2$)$_2$P(C$_2$H$_5$)$_2$.

gous low-spin octahedral complexes at high pressure (Ferraro and Takemoto, 1974; Fisher and Drickamer, 1971; Wood, 1972; Fung and Drickamer, 1969a). Table VIII summarizes some of these results. The initial conversion from the high-spin to the low-spin state has been explained (Drickamer and Frank, 1973; Fisher and Drickamer, 1971) by the increase in ligand-field potential with pressure until it exceeds the electron pairing energy. This initial effect is accompanied by the back donation of the metal $t_{2g}$ electrons into the $\pi^*$ orbitals of the ligand. With a further increase in pressure, this back donation is reduced by the accessibility of electrons from the ligand (Drickamer and Frank, 1973). Pearson (1976) has cited conditions for which an octahedral structure can be converted to a trigonal prism. In certain tris-chelates, in which the symmetry is lower than $O_h(D_3)$, interconversion of isomers may be possible. However, trigonal prism structures are rare.

Adams *et al.* (1981b) examined decacarbonyl dimanganese and decarbonyl-dirhenium with pressure in a DAC and followed the Raman bands in the carbonyl and low-frequency regions. These materials may be considered under a six-coordinate system. Evidence was presented for a conversion of the staggered configuration ($D_{4d}$) for both carbonyls to the eclipsed form ($D_{4h}$). Figure 3 presents the spectral data obtained at ambient pressure and 16 kbar. The Mn(CO)$_{10}$ complex transforms at 8 kbar, and the Re$_2$(CO)$_{10}$ complex at 5 kbar. The authors also report that two further phase transitions occur below 150 kbar for both carbonyls.

## Table VIII

High-Pressure Spin–Spin Interconversions As Diagnosed by Vibrational Spectroscopy[a]

| Compound | Central atom CN | High spin (number of unpaired electrons) | Spin state change | Conversion pressure (kbar)[b] | Experimental probe |
|---|---|---|---|---|---|
| Fe(NO)(salen) ($d^5$) | 5 | 3 | High spin → low spin | 21 | NO stretching region |
| Fe(phen)$_2$I$_2$ ($d^6$) | 6 | 2 | High spin → low spin | partial at ~30 | Skeletal, FIR |
| Fe(phen)$_2$(N$_3$)$_2$ ($d^6$) | 6 | 2 | High spin → low spin | (17)[c] | Skeletal, FIR |
| Fe(phen)$_2$(NCS)$_2$ ($d^6$) | 6 | 2 | High spin → low spin | 18 (8) | Skeletal, FIR |
| Fe(phen)$_2$(NCSe)$_2$ ($d^6$) | 6 | 2 | High spin → low spin | 8–10 (6) | Skeletal, FIR |
| Fe(bipy)$_2$(NCS)$_2$ ($d^6$) | 6 | 2 | High spin → low spin | 15 | Skeletal, FIR |
| Co(nnp)(NCS)$_2$ ($d^7$) | 5 | 3 | High spin → low spin | 4 | Skeletal, FIR |
| [Co(np$_3$)I][B$\phi_4$] ($d^7$) | 5 | 3 | High spin → low spin | 20 | Electronic region |
| Ni(Bz$\phi_2$P$_2$Br$_2$ ($d^8$) | 4 | 2 | High spin → low spin | 12 | Skeletal, FIR, and electronic regions |
| Fe(EPDTC)$_3$ ($d^5$) | 6 | 5 | High spin → low spin | 35 | Skeletal, FIR |

[a] Structural conversion also occurs. Abbreviations: CN = coordination number; phen = phenanthroline; bipy = bipyridyl; 

nnp = Et$_2$N—(CH$_2$)$_2$—N(H)—(CH$_2$)$_2$P$\phi_2$; Bz = benzyl; $\phi$ = phenyl; salen = $NN'$-ethylenebis(salicylideneimine),

[structure of salen ligand shown]

EPDTC = $n$-ethyl-$n$-phenyldithiocarbamato; np$_3$ = tris(2-diphenylphosphinoethyl)amine.
[b] Average pressure except where indicated.
[c] Parentheses indicate hydrostatic pressures [results of Adams et al. (1982)].

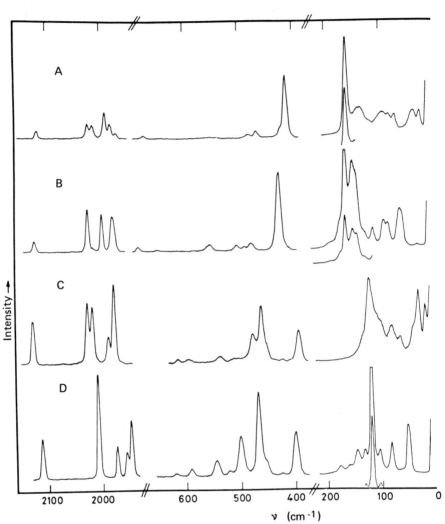

**Fig. 3.** Raman spectra of $Mn_2(CO)_{10}$ at (A) ambient pressure, (B) 16 kbar, and of $Re_2(CO)_{10}$ at (C) ambient pressure, and (D) 16 kbar in diamond anvil cell. [From Adams et al. (1981b).]

# 1. Solid-State Structural Conversions

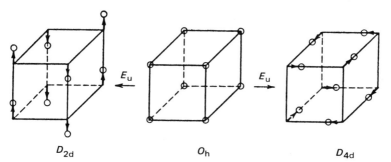

**Fig. 4.** The $E_u$ modes that convert a regular cubic $XY_8$ molecule $O_h$ into a dodecahedron $D_{2d}$ or Archimedian antiprism $D_{4d}$. [From Pearson (1976).]

### e. Seven- and Eight-Coordinate Complexes

Nonrigid configurations for seven- and eight-coordinate complexes have been demonstrated in solution studies (Malatesta, 1964; Fleischer *et al.*, 1972; Hoard and Silverton, 1963; Parish, 1966; Klanberg *et al.*, 1967; Lippard, 1966; Parish and Perkins, 1967; Muetterties, 1973; Claassen *et al.*, 1968; Rossman *et al.*, 1973; Hoard *et al.*, 1968; Hartman and Miller, 1968; McGarvey, 1966; Hayes, 1966). However, no solid-state high-pressure studies have been reported to date.

Interconversion between a $D_{5h}$ pentagonal bipyramid and a $D_{3v}$ capped octahedron has been observed (Burbank and Bartlett, 1968) for a seven-coordinate system. The common structures for coordination number eight are the $D_{2d}$ dodecahedron and the $D_{4d}$ Archimedian antiprism (see Fig. 4). Potential energy calculations for the two models show very little energy difference and no barrier between them (Blight and Kepert, 1968). It would appear that pressure studies involving seven- and eight-coordinate systems should prove worthwhile.

### f. Nonrigidity of Solids at High Pressures

*(1) Summary of Interconversions with Pressure*  It may be concluded that solid-state high-pressure structural transformations are possible in transition-metal complexes. All of the transformations examined thus far have been reversible, with the exception of that in Ni(Qnqn)Cl$_2$, in which a dimer is formed at high pressure. In

this complex, two additional bonds are formed on dimerization, and they contribute to the stability of the high-pressure phase.

The probability of producing structural interconversions with pure or nearly pure tetrahedral and octahedral complexes is predicted to be small on the basis of the theoretical considerations discussed (Pearson, 1969a,b; Bader, 1962). For undistorted tetrahedral complexes, this prediction is borne out by experiment. Attempts (Ferraro, 1975, unpublished data) to convert complexes of nearly tetrahedral symmetry have been unsuccessful. The results of this work, however, indicate that the pressure-induced conversions of tetragonally distorted tetrahedral complexes are possible, in particular, where an asymmetric ligand field is observed by the central atom (e.g., a complex involving several types of ligands). The importance of a distorted structure appears to be a necessity in a solid-state pressure conversion (Ferraro, 1976).

It is possible that distorted six-coordinate complexes may also behave similarly (Eisenberg and Ibers, 1965; Smith *et al.*, 1965; Eisenberg and Gray, 1967). The results observed to date for four- and six-coordinate complexes are not surprising because the energy barrier to rearrangement for true tetrahedral and octahedral structures is certainly high. For the $Ni(BzPh_2P)_2Br_2$ complex, the unpaired electrons may contribute to the lowering of the energy difference between the distorted tetrahedral and the square-planar configurations. This effect would be superimposed upon the beneficial effect of a starting structure that is distorted toward the square-planar geometry.

For five-coordinate complexes the energy barrier for structural interconversion is small, and many examples have been reported in which the trigonal-bipyramidal and square-pyramidal isomers both exist (Wood, 1972; Muetterties, 1970a,b; Holmes, 1972; Muetterties and Schunn, 1966; Muetterties and Wright, 1967). This is apparently also true in the solid state because high-pressure studies indicate that interconversion is readily obtained. In systems containing a tripodlike tetradentate ligand, the ligand flexibility favors the trigonal-bipyramidal structure. For the $Ni(CN)_5^{3-}$ ion, a structure that is distorted in the direction of the pressure-stable pyramidal phase, the monodentate cyanide ion permits the rearrangement to occur.

As a result of this research, we have arrived at a new scheme of classification of the types of behavior observed in transition-metal compounds at high pressure (see Table IX) (Drickamer and Frank,

## 1. Solid-State Structural Conversions

*Table IX*

*Behavior Classes for Pressure-Induced Solid-State Changes*[a]

| Behavior class | Structural change | | Electronic change | | Examples |
|---|---|---|---|---|---|
| | Geometric change | CN change | Spin-state change | Oxidation-state change | |
| 1 | No | No | No | No | Green Ni $(BzPh_2P)_2Cl_2$<br>$[Ni(Qnqn)Cl_2]_2$<br>$Co(Qnqn)Cl_2$<br>$FeS_2$ |
| 2 | Yes | No | Yes | No | Red $Ni(BzPh_2)_2Br_2$,<br>$[Co(np_3)I][B\phi_4]$ |
| 3 | Yes | No | No | No | Several $CuCl_4^{2-}$, $Mn_2(CO)_{10}$,<br>$Re_2(CO)_{10}$<br>$Ni(CN)_5^{3-}$ |
| 4 | Yes | Yes | No | No | $Ni(Qnqn)Cl_2$, $Co(py)_2Cl_2$<br>NaCl<br>Graphite, $\beta$-quartz,<br>$MnF_2$ |
| 5 | No | No | Yes | No | $Fe(phen)_2(N_3)_2$<br>$Fe(phen)_2(NCS)_2$ |
| 6 | No | No | No | Yes | $Fe(acac)_3$<br>$Cu(OXin)_2$ |
| 7 | No | No | Yes | Yes | Hemin |
| 8 | Yes | | | Yes | $Co(NO)(Ph_2CH_3P)_2Cl_2$ |

[a] Abbreviations: CN = coordination number; Bz = benzyl; Qnqn = *trans*-2-(2'-quinolyl)methylene-3-quinuclidione; py = pyridine; acac = acetylacetonate; OXin = 8-hydroxyquinoline.

1973; Long and Ferraro, 1973; Long and Coffen, 1974; Vaughan and Drickamer, 1967; Willett *et al.*, 1974; Wang and Drickamer, 1973; Basile *et al.*, 1974; Fisher and Drickamer, 1971; Ferraro and Takemoto, 1974; Bargeron and Drickamer, 1971; Frank and Drickamer, 1972; Feltham *et al.*, 1979, unpublished data). The eight behavior classes are based primarily on the presence or absence of a structural and/or electronic change in the complex between ambient and high pressure. Class 1 compounds exhibit neither large structural nor electronic changes, but they would include compounds that show small effects, such as slight unit cell contraction, minor crystallographic changes in space group, small changes in crystal-field parameters, and small shifts in charge-transfer bands. Class 2 compounds exhibit significant structural changes with accompanying spin-state changes. Class 3 includes only compounds that undergo a structural change. Class 4 includes complexes that show a structure change in coordination number. Class 5 includes compounds that show only spin-state changes. Class 6 compounds show only oxidation-state changes. Class 7 includes compounds that show spin-state and oxidation-state changes. Class 8 is less definite; both structural and electronic changes are indicated but not entirely delineated.

*(2) Pressure Effects on Spin States* A number of changes of oxidation and spin state caused by pressure have been reported by Drickamer and Frank (1973). In these cases, Mössbauer or absorption spectroscopy was used as the diagnostic tool to observe the changes. For changes in solutions, see Sinn (1974).

Several systems have been examined by using vibrational spectroscopy as the diagnostic tool to identify spin-state equilibria in the solid state. In these examples, the FIR region, and in particular the metal–ligand vibration, was followed with pressure to determine a change from a high-spin to a low-spin state. The high-spin, metal–ligand vibration occurs at lower energy than that with low spin. In some instances, low temperature are used with pressure to facilitate the conversion.

The complex $NiBr_2(Bz\phi_2P)_2$ was converted at 12 kbar from a high-spin, distorted tetrahedral molecule to a low-spin, square-planar configuration (Ferraro *et al.*, 1973). In this complex a structural change as well as a spin-state conversion occurred. $Co(nnp)(NCS)_2$, where nnp represents $Et_2N-(CH_2)_2-NH-(CH_2)P\phi_2$, was converted at 10 kbar and 150 K from the high-spin to the low-spin state (Sacconi and Ferraro, 1974). The relationship of the equilibria with temperature and pressure is shown below.

```
10 kbar ┬ LS    LS + HS    HS
 1 atm  ┴ LS    HS
        └─────────────────────
          100    RT    353    T (K)
```

The $Fe(phen)_2X_2$ and $Fe(bipy)_2X_2$ complexes—where phen represents 1,10-phenanthroline, bipy represents 2,2'-bipyridine, and X represents $Cl^-$, $Br^-$, $N_3^-$, $NCO^-$, $OAc^-$, and $HCOO^-$—exist in high-spin states (Ferraro and Takemoto, 1974). The complexes Fe(phen)$_2$(NCS)$_2$, Fe(phen)$_2$(NCSe)$_2$, and Fe(bipy)$_2$(NCS)$_2$ were studied at high pressure and/or low temperature. Complete conversion with pressure to low spin did not occur. However, the mixtures of high-spin and low-spin forms maintained at high pressures could be converted to the low-spin state if the samples were cooled to 100 K. The high-spin state can be converted to the low-spin state directly with cooling to 100 K. The results parallel those obtained by Fisher and Drickamer (1971), who used Mössbauer techniques for Fe(phen)$_2$(NCS)$_2$ and Fe(phen)$_2$(NCSe)$_2$. In these studies the Fe–N(phen) vibration was followed with pressure. This vibration appears at lower frequency in the high-spin state and shifts to higher frequency with pressure in the low-spin state.

Using hydrostatic conditions in the diamond cell, Adams et al. (1982) repeated some of these studies. Their results served to illustrate that nonhydrostatic conditions that are subjected to shear stresses may tend to counteract the conversion of high spin to low spin. For Fe(o-phen)$_2$(NCS)$_2$ and Fe(o-phen)$_2$(NCSe)$_2$, the spin-state transitions occurred at 8 and 6 kbar, respectively. In the earlier nonhydrostatic work the conversions occurred at 18 and 8–10 kbar, respectively. For Fe(o-phen)$_2$(N$_3$)$_2$, the transition occurred at 17 kbar, and for Fe(o-phen)I$_2$, the conversion to low spin increases with pressure but up to 30 kbar never converts totally.

High-spin–low-spin crossovers with pressure have also been observed for tri(N-ethyl-N-phenyldithiocarbamato)iron(III) (Butcher et al., 1976a,b) and represent the first pressure conversions of Fe(III) to low spin. These results may have some consequences in problems relating to the earth's mantle.

The results of the low-temperature conversion to low spin can be explained in terms of a strengthening of the Fe–N(phen or bipy) and Fe–N(NCS and NCSe) bonds due to back donation of the $t_{2g}$ electrons of the metal to the $\pi^*$ orbitals of the organic ligand and NCSe. This mechanism may also be present at the outset of pressure application, but the back donation of the metal is reduced with increasing pressure by the accessibility of the $\pi$ electrons from the ligand $\pi^*$ orbitals (Drickamer and Frank, 1973).

The five-coordinate square-pyramidal complex Fe(NO)(salen), where salen is $N,N'$-ethylene bis(salicylidenimine), has been shown to contain iron in an intermediate spin state ($S = \frac{3}{2}$) and to exhibit spin equilibrium at low temperature (Earnshaw et al., 1969). Mössbauer results have been analyzed on the basis that the complex contains Fe(III) and $NO^-$. In a study under pressure, Feltham *et al.* (1979) shows a shift of the NO vibrational band toward lower frequency (from 1710 to 1630 $cm^{-1}$) with a pressure increase or with cooling (Haller *et al.*, 1979) (see Fig. 5). These results can be consistent with a spin-state change for Fe to $S = \frac{1}{2}$ that occurs with a cooling or a pressure increase.

Table VIII summarizes high-pressure spin-state conversions as diagnosed by vibrational spectroscopy.

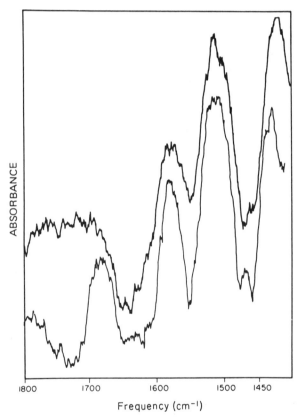

*Fig. 5.* Infrared spectra of solid Fe(NO)(salen) at ambient pressure (lower curve) and at 37 kbar (upper curve). The experiments were carried out in a DAC. [From Haller *et al.* (1979).]

# 1. Solid-State Structural Conversions

Ferraro et al. (1979) examined the systems $(Co(np_3)X)(B\phi_4)$, where X represents Cl, Br, and I under pressure. The electronic region was used as the probe, since overlapping absorption precluded making FIR studies in the Co–N, Co–P region ($<400$ cm$^{-1}$) and of the Co–Br, CoI vibrations at $<250$ cm$^{-1}$. Figure 6 shows the

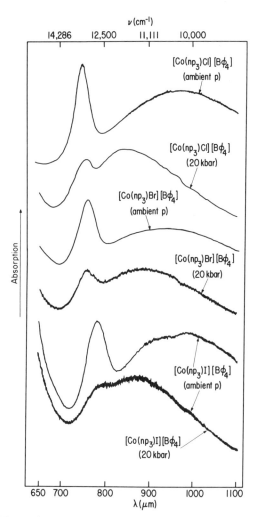

**Fig. 6.** Electronic spectra of [Co(np$_3$)Cl][B$\phi_4$], [Co(np$_3$)Br][B$\phi_4$], [Co(np$_3$)I][B$\phi_4$] at ambient and 20-kbar pressure. [From Ferraro et al. (1979).]

results obtained. The complexes were examined at ambient pressure and the spectra compared with spectra obtained at 20 kbar. It can be observed that the 13,000-cm$^{-1}$ absorption disappears in the (Co(np$_3$)I)(B$\phi_4$) case and is indicative of a crossover to a low spin state. However, in the case of the chloro or bromo analog, some high-spin states remain up to 35 kbar. The tendency to pair spins appears to be greatest for the iodine complex and may account for the fact that both high- and low-spin states are found at room temperature for this complex and not for the chloro or bromo analogs.

Corollary data were obtained in the low-frequency vibrational spectrum of (Co(np$_3$)Cl)B$\phi_4$. An absorption at 300 cm$^{-1}$, assigned to a cobalt–chlorine stretching vibration in the high-spin state, weakens in intensity with pressure, and an intensification of absorption occurs at 350 cm$^{-1}$. This is the direction of the frequency shift that is observed for the metal–ligand vibration in going from high spin to low spin. Overlapping ligand absorptions precluded studying the Co–N, Co–P vibrations in the region <400 cm$^{-1}$ and Co–Br, Co–I vibrations in the region <250 cm$^{-1}$.

It can be concluded that high pressure favors the low-spin state and the tetragonal pyramidal geometry in five-coordinate complexes of the type (Co(np$_3$)X)(B$\phi_4$). This system illustrates the behavior class type in which geometrical and spin-state changes occur with pressure but the coordination number and oxidation state remain unchanged (Ferraro and Long, 1975). The results parallel those obtained for (Cr(en)$_3$)(Ni(CN)$_5$), where the geometry of the Ni(CN)$_5^{3-}$ changes from trigonal-bipyramidal to square-pyramidal with an increase in pressure (Basile *et al.*, 1974), and for Co(nnp)(NCS)$_2$ for which the low-spin state is favored (Sacconi and Ferraro, 1974).

*(3) Pressure Oxidation–Reduction* Hellner *et al.* (1974) observed a pressure-dependent reduction of Fe(III) to Fe(II) that occurred in (N(CH$_3$)$_4$)$_3$(Fe(NCS)$_6$). The reduction was characterized by the appearance of a new band at 238 cm$^{-1}$, lower in frequency than the 295, 270-cm$^{-1}$ vibrations attributed to Fe(III)–N mode. Figure 7 presents the results in the far-infrared region. The study was made in an anvil cell and confirmed Mössbauer studies by Fung and Drickamer (1969b). No evidence for ligand isomerization was found. A similar pressure reduction was reported for (Mn(III)(CN)$_6$)$^{3-}$ and (Fe(III)(CN)$_6$)$^3$ by Hellner *et al.* (1973). They followed the CN mode with pressure. On release of pressure, a slow reoxidation ensued.

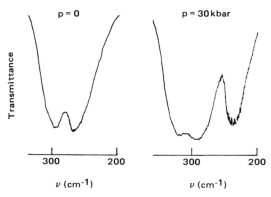

**Fig. 7.** Infrared spectra in the $\nu_{FeN}$ region for $[N(CH_3)_4]_3[Fe(NCS)_6]$ at ambient pressure and at 30 kbar. [From Hellner et al. (1974).]

Two isomers of $Co(NO)(Ph_2CH_3P)_2Cl_2$ have been reported by Brock et al. (1973). One isomer is trigonal-bipyramidal (TBP) and contains Co(I) and (NO$^+$) ions, most probably with a linear Co–N–O bond. The second isomer is square-pyramidal and has Co(III) and (NO$^-$) ions with a bent Co–N–O bond. The NO vibration occurs at 1750 cm$^{-1}$ in the TBP complex and at 1650 cm$^{-1}$ in the square-pyramidal complex. With an increase in pressure the TBP complex appears to convert to the square-pyramidal geometry (Feltham et al., 1979, unpublished data).

For other oxidation–reduction effects of pressure, see Drickamer and Frank (1973), who used Mössbauer as the diagnostic technique.

*(4) Ligand Isomerism* Ligand isomerism has been characterized for some time. In most cases, the isomerism has taken place in solution. Very little information has become available relative to this phenomenon occurring in the solid state.

Nitro and nitrito complexes are of interest in this discussion. The $NO_2$ group coordinates to a metal in a variety of ways, including the following.

Nitro complex I

Nitro complex II

Chelating nitro complex III

```
    M        M            M   M           M—O     O—M
     \      /              \ /              \\   //
      N—O                   O                 N
     ‖                      |                 
     O                      N                VI
     IV                    ‖ \\
                           O   O
                              V
```

Bridging nitro complexes

A study of solid $(Ni(en)_2(NO_2)_2)$ at pressure of 30 kbar by Ferraro and Fabbrizzi (1978) has demonstrated that the violet form (obtained at 120°C) was transformed to the red form with pressure. The red isomer is the nitro form, while the violet isomer is the nitrito form. The 560-nm band in the visible was used to monitor the transformation.

*(5) High-Pressure Effects on Vibrational Transitions* As further research involving high-pressure effects coordination has continued, it has become obvious that certain vibrations in solid materials are more sensitive to external pressure than are others. Vibrations that involve expansion of molecular volume appear to be quite pressure sensitive. The factors contributing to pressure effects on vibrations are complicated, since many of them contribute to the pressure sensitivity. For solids such factors are packing effects, compressibility, type of bond (ionic versus covalent), crystal field stabilization energy (CFSE), steric effects, electronic repulsions, electronic delocalization effects, geometry of molecule, and other effects. All contribute to the determination of the pressure sensitivity of vibrations in a molecule. The pressure sensitivity of vibrations has suggested that the technique can be useful along with metal isotope studies to aid in making far-infrared assignments in coordination compounds (Nakamoto *et al.*, 1970).

Several examples of pressure effects on vibrations for different coordination complexes will be discussed in the ensuing sections.

(a) *Tetrahedral Complexes* $MX_4^{2-}$ * Complex halides such as $K_2PtCl_4$ and $K_2PdCl_4$ were studied by Ferraro (1970) to 33-kbar average pressure. Table X tabulates the results. The $E_u$ internal mode was found to be more pressure sensitive than the $A_2$ internal mode, which tends to disappear with a pressure increase. The most pressure sensitive external band was the translational mode along the

---

* $MX_4^{2-}$, where X represents Cl and CN.

## 1. Solid-State Structural Conversions

### Table X
*Pressure Dependencies for $K_2PtCl_4$ and $K_2PdCl_4$*

| Compound | Vibration | Ambient pressure $\nu$ (cm$^{-1}$) | Pressure dependence[a] $d\nu/dp$ (cm$^{-1}$ kbar$^{-1}$) |
|---|---|---|---|
| $K_2PtCl_4$ | $\nu_{PtCl}$ ($E_u$) | 325 | 0.25 |
| | $\delta_{ClPtCl}$ ($E_u$) (in plane) | 195 | 0.46 |
| | $\delta_{ClPtCl}$ ($A_{2u}$) (out of plane) | 173 | Disappears[b] |
| | Lattice ($E_u$—$a_0$ axis) | 116 | 0.86 |
| | Lattice ($A_{2u}$—$c_0$ axis) | 106 | 0.08 |
| | Lattice ($E_u$—$b_0$ axis) | 89 | Disappears[c] |
| $K_2PdCl_4$ | $\nu_{PdCl}$ ($E_u$) | 340, 337 | 0.42 (Coalesces into one band) |
| | $\delta_{ClPdCl}$ ($E_u$) (in plane) | 190 | 0.35 |
| | $\delta_{ClPdCl}$ ($A_{2u}$) (out of plane) | 170 | Disappears[b] |
| | Lattice ($E_u$—$a_0$ axis) | 126 | 0.64 |
| | Lattice ($A_{2u}$—$c_0$ axis) | 115 | 0.15 |
| | Lattice ($E_u$—$b_0$ axis) | 96 | Disappears[b] |

[a] Where a frequency shift occurs it is toward higher energy (blue shift) $d\nu = (\nu_{33 \text{ kbar}} - \nu_{0.001 \text{ kbar}}) = dp = (p_{33 \text{ kbar}} - p_{0.001 \text{ kbar}})$.
[b] Band decreases in intensity and disappears.
[c] Band decreases in intensity and shifts into stationary $A_{2u}$ band.

long axis $a_0$, while the mode along the short axis $c_0$ was less pressure dependent. It was concluded that contractions occurred along the $a_0$ axis but that a contraction or even an expansion occurred along the $c_0$ axis with a pressure increase.

(b) **Complexes $K_2M(CN)_4$, $M(CN)_2$, $M(CN)_2^-$*** Complex cyanides of the type $K_2M(CN)_4$ and $M(CN)_2$, where M represents Zn, Cd, Hg, and Ni have been studied by pressure to 30 kbar and the effects followed by infrared spectroscopy (Dehnicke *et al.*, 1974; Hellner *et al.*, 1973). The CN stretching region as well as the M–CN region (200–600 cm$^{-1}$) were followed. For the $M(CN)_4^{2-}$ tetrahedra, it was found that the $F_2$ vibrations lost degeneracy with pressure. For Cd or Zn, the symmetry lowered to $D_{2d}$ or $C_{3v}$. The symmetry

* Where M represents Zn, Cd, Hg, and Ag.

for $Zn(CN)_2$ was found to lower while that of $Cd(CN)_2$ and $Hg(CN)_2$ increased with pressure. Complementary Raman studies for $K_2M(CN)_4$ (where M represents Zn, Cd, and Hg) were made under pressure by Adams et al. (1981c). Each complex manifested two first-order phase transitions with pressure. These occur at 1.5 and 8.5 kbar for Hg, 3 and 8 kbar for Cd, and 4 and 14 kbar for Zn. It was concluded that for each complex, phase II has the trigonally distorted spinel structure at room temperature of $Rb_2Hg(CN)_4$. Phase III was found to possess the hausmanite-type structure (tetragonally distorted spinel).

Wong (1979) studied the pressure-induced phase transitions and pressure dependency of the lattice modes in $K[Ag(CN)_2]$ by Raman scattering. Two high-pressure phases were found below 18 kbar. The pressure dependencies for the various vibrations in the three phases are recorded in Table XI. The CN and AgC stretching vibrations are lowered in the higher pressure phases, indicative of a higher coordination number. The rotational lattice modes were strongly pressure dependent. All lattice modes shift blue with pressure. The lattice mode at ~68 cm$^{-1}$ in phase II is found to be softened by pressure and has a negative pressure coefficient.

(c) Square-Planar Complexes $MX_2L_2$ Allkins et al. (1969) have studied square-planar complexes of the type $MX_2L_2$, where M represents $Pt^{2+}$ or $Pd^{2+}$, X represents Cl or Br, L represents $(CH_3)_2S$, at pressures of up to 50 kbar. Infrared active vibrations were monitored in a DAC. Generally, vibrations involving expansion of molecular volume were more pressure-sensitive. For these complexes, the vibrations M–X symmetric stretch, C–S–C symmetric bend, and C–H symmetric stretch were the most significantly affected by pressure, both in frequency shifts and in intensity changes. The results are parallel to those of Postmus et al. (1967) for symmetric M–X stretching modes in various complexes containing M–X stretching modes. The latter results are discussed in a later section.

(d) Octahedral Complexes $M(NH_3)_6^{2+}$, $M(NH_3)_6^{3+}$, $MX_6^{2-}$, $MX_6$ High-pressure studies (to 40 kbar) on complexes of the type $(Ni(NH_3)_6)X_2$ and $(Co(NH_3)_6)X_3$, were X represents Cl, Br, and I, have been made; and the far infrared has been used to diagnose the changes occurring (Adams and Payne, 1976). The Ni(II) complexes crystallize in a cubic (fluoride) structure. The selection rules for

## 1. Solid-State Structural Conversions

*Table XI*

*Pressure Dependencies and Assignments of the Raman Bands of* $K[Ag(CN)_2]$

| Phase I | | Phase II | | Phase III | | |
|---|---|---|---|---|---|---|
| $\nu$ (cm$^{-1}$) (1 atm) | $d\nu/dp$ (cm$^{-1}$ kbar$^{-1}$) | $\nu$ (cm$^{-1}$) (2.4 kbar) | $d\nu/dp$ (cm$^{-1}$ kbar$^{-1}$) | $\nu$ (cm$^{-1}$) (16.8 kbar) | $d\nu/dp$ (cm$^{-1}$ kbar$^{-1}$) | Assignments |
| 2147.1 | 0.71 | 2139.2 | −0.17 | 2142.3 | 0.20 | $\nu_1$ |
| — | — | — | — | 2133.1 | −0.09 | |
| 363.7 | — | 359.0 | — | 356.0 | 0.65 | $\nu_2$ |
| — | — | — | — | 323.4 | 0.80 | |
| 331.2 | — | — | — | 293.9 | 0.17 | $\nu_6$ |
| 249.72 | 0.40 | 257.6 | 0.55 | 269.6 | 0.37 | $\nu_5$ |
| — | — | — | — | 258.7 | 0.20 | |
| 140.0 | — | 143.7 | — | 177.3 | — | $\nu_3$ |
| 105.9 | 4.90 | 119.2 | 1.78 | 160.4 | 1.15 | |
| 93.1 | — | 99 | 1.36 | 150.4 | 1.35 | |
| 77.2 | — | 80.2 | 1.91 | 139.2 | 0.99 | |
| 66.2 | 3.37 | 68.0 | −0.02 | 125.8 | — | |
| 53.7 | 3.36 | 56.3 | 0.37 | 121.9 | — | |
| 41.8 | — | 23.4 | — | 115.5 | 0.68 | Lattice |
| 29.6 | — | 13.8 | — | 97.8 | 0.66 | |
| 23.6 | — | — | — | 68.9 | 0.11 | |
| 14.3 | — | — | — | 60.6 | 0.42 | |
| | | | | 38.0 | 0.60 | |
| | | | | 27.2 | 0.20 | |

such a structure predict two active far-infrared-active $\nu_3$ and $\nu_4$ vibrations and a lattice mode, all belonging to an $F_{1u}$ representation. Table XII summarizes the data. For the Ni(II) complexes a phase change with a lower symmetry was indicated. Both $\nu_3$ and $\nu_4$ were found to split or broaden considerably. $(Co(NH_3)_6)I_3$ has an $O_h^5$ symmetry, and two $F_{1u}$ lattice modes are predicted, along with two $F_{1u}$ internal modes. Both $\nu_4$ and $\nu_L$ showed considerable blue shifts with pressure.

Complex halides of the type $M_2M'Cl_6$, where M represents K, Rb, Cs, Tl, or $NH_4$ and M' represents Pt or Pd, were subjected to pressures and the pressure sensitivities of internal and lattice modes followed. Both Raman and infrared data were used to follow the changes occurring. Early infrared work by Ferraro (1970) on $K_2PtCl_6$ and $K_2PdCl_6$ followed the two $F_{1u}$ lattice vibration with

## Table XII

Infrared Vibrational Wave Numbers ($Cm^{-1}$) for Some Hexa-Ammine Complexes at Various Temperatures and Pressures

| Complex | | 0.001 kbar/77 K | | | 0.001 kbar/290 K | | | | 40 kbar/290 K | | |
|---|---|---|---|---|---|---|---|---|---|---|---|
| | | $v_3$ | $v_4$ | $v_L$ | $v_3$ | $v_4$ | $v_L$ | | $v_3$ | $v_4$ | $v_L$ |
| [Ni(NH$_3$)$_6$]X$_2$ | X = Cl | 344 | 219[a] | 120 | 335 | 216 | 115 | | 342,370 | 226 | 138 |
| | X = Br | — | — | — | 329 | 218 | 91 | | 334,368 | 225,265 | 107 |
| | X = I | 329 | 219 | 83[b] | 322 | 215 | 82 | | 332 | 258,217 | 100 |
| [Co(NH$_3$)$_6$]X$_3$ | X = Cl | — | 269,335 | 94,156 | — | 332 | 148 | | — | 341 | 175 |
| | X = Br | — | — | — | — | 320 | 118 | | — | 325 | 144 |
| | X = I | — | 320 | 88,109[c] | — | 318 | 101[d] | | — | 330 | 112,140 |

[a] Approximate value only.
[b] There is also a strong, very broad band ~150 $cm^{-1}$, attributed to $\tau(NH_3)$.
[c] Weak, very broad absorption ~200 $cm^{-1}$, probably due to $\tau(NH_3)$.
[d] Doublet with unresolved components ~86,106 $cm^{-1}$.

## Table XIII

Vibrational Frequencies ($Cm^{-1}$) at Ambient Pressure and Pressure Sensitivities $\Delta v/\Delta p$ ($cm^{-1}$ $kbar^{-1}$) for Solids $M_2[PtCl_6]$

| Solid | Raman | | | | | | | I.r. | | | | |
|---|---|---|---|---|---|---|---|---|---|---|---|---|
| | $v_1$ | $\Delta v/\Delta p$ | $v_2$ | $\Delta v/\Delta p$ | $v_5$ | $\Delta v/\Delta p$ | | $v_3$ | $\Delta v/\Delta p$ | $v_4$ | $v_L$ | $\Delta v/\Delta p$ |
| K$_2$[PtCl$_6$] | 351 | 0.85 | 320 | 0.7 | | | | 345 | 0.4 | 187 | 88 | 0.9 |
| Rb$_2$[PtCl$_6$] | 345 | 1.45 | 316 | 1.05 | | | | 342 | 0.4 | 187 | 78 | 1.0 |
| Cs$_2$[PtCl$_6$] | 337 | 0.8 | 311 | 0.5 | | | | 332 | 0.35 | 186 | 70 | 0.9 |
| (NH$_4$)$_2$[PtCl$_6$] | 346 | 1.15 | 316 | 0.9 | | | | 340 | 0.5 | 201 | 135 | 0.85 |
| Tl$_2$[PtCl$_6$] | 347 | 1.6 | 317 | 1.1 | 163 | 1.2 | | 335 | 0.65 | 181 | 49 | 0.6 |

pressures of up to ~33-kbar average pressure. Later Raman studies by Adams and Payne (1975) for $M_2PtCl_6$ complexes with various M cations as indicated earlier showed that the (Pt–Cl) modes' sensitivity is in the order $\nu_1(A_{1g}) > \nu_2(E_g) > \nu_3(F_{1u})$ at pressures of up to ~20 kbar. For $\nu_1$ and $\nu_2$, the order for the various complexes was Cs < K < $NH_4$ < Rb < Tl. The $\nu_3$ vibration followed the same order, but the shifts were smaller. The complex $Tl_2(PtCl_6)$ showed a change in slope in the plot of $\nu$ versus $p$, and this was interpreted as being due to a phase change. Table XIII summarizes the results of Adams and Payne (1975). Adams *et al.* (1981a) extended the work involving $K_2PtCl_6$ to hydrostatic conditions. When the hydrostatic results are compared with earlier work (Ferraro, 1970; Adams and Payne, 1975) under nonhydrostatic pressures, the nonhydrostatic pressure results underestimate the true shifts of the infrared modes.

A far-infrared study of salts of $SeX_6^{2-}$, where X represents Cl or Br, and $TeX_6^{2-}$, where X represents Cl, Br, and I, at pressures up to ~40 kbar was made by Adams *et al.* (1976). All observed bands shifted toward the blue with the $\nu_4$ bending mode exhibiting the most dramatic effect. The $\nu_4$ vibration is not present for some compounds but appears with increasing pressure. Other compounds showing $\nu_4$ at ambient pressures lose the vibration with increasing pressure. The results were explained on the basis of an anion–cation interaction, which allows a pressure-dependent delocalization of the inert-pair electrons in the lattice of the salts. Table XIV tabulates the results of the study.

Adams *et al.* (1981a) conducted a far-infrared and Raman study of complex chlorides of the type $A_2MCl_6$, where A represents Cs, Rb, or K and M represents Sn, Te, or Pt, in the DAC under hydrostatic conditions. By using compressibilities calculated from unit cell constants and lattice energies, Grüneisen parameters were obtained. These parameters were found to be greater for $K_2SnCl_6$ and $Rb_2TeCl_6$ than for transition metal salts.

Breitinger and Kristen (1977) conducted Raman studies under pressures of up to 25 kbar of $TlSbF_6$ and $BaMF_6$, where M represents Si, Ge, Sn, and Ti. The space group for these materials is $D_{3d}^5$ ($Z' = 1$). The $E_g$ libration in the Raman spectrum demonstrates a higher pressure sensitivity than do the internal modes of the $MF_6^{2-}$ anions. Table XV summarizes the results for the various materials studied.

Hellner *et al.* (1973) conducted an infrared study of the com-

## Table XIV
### Pressure Dependencies of the IR-Active Vibrations of Some Inert Pair Hexahalides, $A_2[MX_6]$

| | $\nu_3$ | | | $\nu_4$ | | | $\nu_L$ | | |
|---|---|---|---|---|---|---|---|---|---|
| | 0.001 kbar | 40 kbar[a] | $\Delta\nu/\Delta p$[b] | 0.001 kbar | 40 kbar[a] | $\Delta\nu/\Delta p$[b] | 0.001 kbar | 40 kbar[a] | $\Delta\nu/\Delta p$[b] |
| $K_2[TeCl_6]$[c] | 197 | 208 | 0.55 | | | | 65 | 85 | 1.00 |
| $Rb_2[TeCl_6]$ | 262 | 278 | 0.80 | 148[d] | 159 | 1.10 | 66 | 82 | 0.80 |
| $Cs_2[TeCl_6]$ | 256 | 265 | 0.45 | 135 | 164 | 1.45 | 110 | 130 | 1.00 |
| $[NH_4]_2[TeCl_6]$ | 255 | 265 | 0.50 | 151 | 178[d] | 2.70 | 78 | e | e |
| $[NMe_4]_2[TeCl_6]$ | 236 | 241 | 0.25 | | | | | | |
| $[NBu_4]_2[TeCl_6]$ | 255 | 261 | 0.30 | | | | | | |
| $K_2[TeBr_6]$[c] | 200 | 208 | 0.40 | 105[d] | 125[d] | 2.00 | 77, 53 | f | |
| $Rb_2[TeBr_6]$ | 201 | 212 | 0.55 | 107[d] | 115 | 0.80 | 54 | 72 | 0.90 |
| $Cs_2[TeBr_6]$ | 200 | 210 | 0.50 | 103[d] | 115 | 1.20 | 55 | 73 | 0.90 |
| $[NH_4]_2[TeBr_6]$ | 195 | 205 | 0.50 | 125 | 147[g] | 1.40 | 82 | 91[g] | 0.60 |
| $[NEt_4]_2[TeBr_6]$ | 176 | 184 | 0.40 | | | | 63 | e | e |
| $[PyH]_2[TeBr_6]$ | 174 | 179 | 0.25 | | | | 60 | e | e |
| $K_2[TeI_6]$[h] | 160 | 165 | 0.25 | 98[i] | 111[i] | 1.30 | 63 | e | e |
| $Cs_2[TeI_6]$ | 159 | 167 | 0.40 | 88 | 96 | 0.40 | 45 | 58[g] | 0.85 |
| $[NH_4]_2[TeI_6]$ | 158 | 168 | 0.50 | | | | | | |
| $[NEt_4]_2[TeI_6]$ | 157[j] | 164 | 0.35 | 100 | 107 | 0.35 | 55 | 66 | 0.55 |

| | | | | | | | | |
|---|---|---|---|---|---|---|---|---|
| [PyH]$_2$[TeI$_6$] | 147 | 151 | 0.20 | 103 | 109$^k$ | 1.20 | 50 | 57$^g$ | 0.45 |
| Rb$_2$[SeCl$_6$] | 285 | 296 | 0.55 | 157 | $e$ | $e$ | 75 | 98 | 1.15 |
| Cs$_2$[SeCl$_6$] | 278 | 285 | 0.35 | 152 | $e$ | $e$ | 73 | 94 | 1.05 |
| [NH$_4$]$_2$[SeCl$_6$] | $l$ | $l$ | $l$ | $l$ | $l$ | $l$ | 132 | 152 | 1.00 |
| K$_2$[SeBr$_6$] | 231 | 239 | 0.40 | 132$^i$ | 147 | 1.00 | 82 | 98 | 0.80 |
| Rb$_2$[SeBr$_6$] | 232 | 243 | 0.55 | 103,$^m$ 87$^m$ | $m$ | $m$ | 63 | 80 | 0.85 |
| Cs$_2$[SeBr$_6$] | 225 | 232 | 0.35 | 102,$^m$ 87$^m$ | $m$ | $m$ | 63 | 77 | 0.70 |
| [NH$_4$]$_2$[SeBr$_6$] | 227 | 235 | 0.35 | 133 | 184 | 2.50 | 97 | 108 | 0.55 |

$^a$ Pressure at the center of the anvil face.
$^b$ Calculated from the value of the average pressure across the anvil face.
$^c$ Monoclinic at room temperature and pressure.
$^d$ At central pressure of 20 kbar.
$^e$ This band lost intensity under pressure without shifting appreciably.
$^f$ Shifts could not be measured accurately.
$^g$ At central pressure of 30 kbar.
$^h$ Just triclinic at room temperature.
$^i$ Possibly lattice mode of distorted structure.
$^j$ $\nu_3$ is a triplet; position and shift of band at half-height are given.
$^k$ At central pressure of 10 kbar.
$^l$ At ambient pressure, $\nu_3$ and $\nu_4$ are merged into one broad band and $\nu$ is a shoulder; at elevated pressure intensity in the region, $\nu_4$ drops but the merged band shifts rapidly and $\nu_L$ separates clearly.
$^m$ Very weak absorptions that disappear with increasing pressure.

### Table XV

*Frequencies in Some Hexafluorometallate Compounds at Ambient Pressure and Temperature, Pressure Dependencies, and Assignments of Raman Bands*[a]

| Compound | $\nu_{Lib}$ ($E_g$) | $\nu_5$ ($A_{1g}$) | $\nu_5$ ($E_g$) | $\nu_2$ ($E_g$) | $\nu_1$ ($A_{1g}$) | |
|---|---|---|---|---|---|---|
| | 127.7 | | 408.9 | 481.5 | 675.4 | $\nu$[b] |
| BaSiF$_6$ | 0.95 | | 0.25 | 0.53 | 0.45 | $d\nu/dp$[c] |
| | 128.3 | | 338.5 | 488.3 | 640.0 | $\nu$ |
| BaGeF$_6$[d] | 0.50 | | 0.18 | 0.17 | e | $d\nu/dp$ |
| | (1.4, 3.1) | | (0.48) | (0.55) | | |
| | 126.8 | | 299.7 | 484.0 | 626.2 | $\nu$ |
| BaTiF$_6$ | 1.04 | | 0.39 | e | 0.28 | $d\nu/dp$ |
| | 134.1 | 258.9 | 277.9 | 484.7 | 595.2 | $\nu$ |
| BaSnF$_6$ | 1.07 | 0.15 | 0.46 | 0.41 | 0.30 | $d\nu/dp$ |
| | 69.3 | 275.3 | 288.1 | 566.0 | 647.4 | $\nu$ |
| TiSbF$_6$ | 1.91 | 0.20 | 0.51 | 0.42 | 0.28 | $d\nu/dp$ |

[a] $\lambda$ = 488.0 nm. Slit width = 5 and 10 cm$^{-1}$.
[b] Frequency at ambient pressure and temperature (cm$^{-1}$).
[c] Pressure dependence (cm$^{-1}$ kbar$^{-1}$).
[d] In parentheses pressure dependencies for the high-pressure modification, stable at pressures higher than 9.6 kbar.
[e] Overlapped by Raman bands of sapphire.

pounds K$_3$(Cr(CN)$_6$), K$_3$(Mn(CN)$_6$), K$_4$(Mn(CN)$_6$) · 3H$_2$O, K$_3$(Fe(CN)$_6$), K$_4$(Fe(CN)$_6$) · 3H$_2$O, and K$_3$(Co(CN)$_6$) in a DAC with pressures of up to ~31 kbar. The CN stretching region (2000–2200 cm$^{-1}$) and the metal–carbon stretching and bending vibrations (330–600 cm$^{-1}$) were followed. Blue shifts with pressure are noted for all of the vibrations. For the compounds (K$_3$(Mn(CN)$_6$) and K$_3$(Fe(CN)$_6$), a partial reduction of the oxidation state from III to II was claimed to occur with pressure and is discussed in Section 3. Table XVI summarizes the results.

Adams *et al.* (1973) studied W(CO)$_6$ under pressure (to 40 kbar) in a diamond cell and measured the Raman spectra. The compound shows $\nu_{CO}$ bands at 2117 cm$^{-1}$ ($A_{1g}$) and 1998 cm$^{-1}$ ($E_g$) at ambient pressure. With an increase in pressure, the $A_{1g}$ mode shifts toward higher energy to 2130 cm$^{-1}$ at 40 kbar, while the $E_g$ mode moves only 2.5 cm$^{-1}$. The W–CO stretching vibration at 435 cm$^{-1}$ at ambient pressure shifts to higher frequency and is broadened. The pressure-sensitive vibrations are other examples of vibration demonstrating expansion or contraction on movement.

(e) Metal Sandwich Compounds   Nakamoto et al. (1970) studied several metal sandwich compounds under pressure to 35 kbar in a DAC. The infrared-active skeletal modes such as the (M-ring)$_{asym}$ vibration was more pressure sensitive than the ring tilt vibration. Table XVII tabulates the effects.

(f) Complexes Demonstrating Asymmetric and Symmetric Stretching Modes   In complexes for which two or more halogen atoms are bonded to a central atom, two stretching vibrations involving the central atom with the halogen atoms are infrared active. These correspond to the asymmetric and symmetric stretching modes. For a series of compounds in which two stretching modes are observed, pressure application up to ~50 kbar demonstrated different responses (see Postmus et al., 1967). Only minor shifts were observed; however, the symmetric mode showed a large decrease in intensity. Figure 8 illustrates results with ($\phi_4$As)(GeCl$_3$). Similar results were reported by Allkins et al. (1969) for square-planar complexes involving symmetrical MX and CH stretching vibrations. These effects involve vibrations that show expansion of molecular volume during the vibration. It is possible that much of this effect depends on the particular complex, its crystal packing, and the freedom of the atoms from any steric effects and may be of limited usefulness. Nevertheless, the technique may be useful in assigning asymmetric versus symmetric stretching vibration in certain cases.

(g) Complexes with Metal Halide Bridging Vibrations (CoL$_2$X$_2$)   Postmus et al. (1967) have examined complexes that are known to exist in polymeric (lilac) and monomeric (blue or green) forms. The assignments for bridged versus terminal metal-halide stretching vibrations are difficult to make in the FIR. Pressure studies of complexes of the type CoL$_2$X$_2$, where L represents py, 4-Cl(py), or 4-Br(py) and X represents Cl or Br, were made in a DAC with pressures of up to 28 kbar in the far-infrared region. The bridging $\nu$CoX manifested appreciable blue shifts as opposed to the termi-

## Table XVI
### Effects of Pressure on Infrared-Active Modes in Cyanocomplexes of 3d-Elements[a]

| Complex | Vibrational type | | Ambient pressure | 31-Kbar pressure | Ambient pressure after pressure release |
|---|---|---|---|---|---|
| $K_3[Cr(CN)_6]$ | $\nu_{CN}$ | $(F_{1u})$ | 2125 (st) | 2132 (st) | 2123 (st) |
| | $\nu_{CrC}$ | $(F_{1u})$ | 330 (vvs, br) | 334 (vvs, br), 343 (sh) | 331 (vvs, br) |
| | $\delta_{CrCN}$ | $(F_{1u})$ | 445 (vvs) | 445 (vvs) | 447 (vvs) |
| $K_3[Mn(CN)_6]$ | $\nu_{CN}$ | $(F_{1u})$ | 2105 (st) | 2116 (st), 2085 (m–st) | 2103 (st), 2058 (s–m) |
| | $\nu_{MnC}$ | $(F_{1u})$ | 352 (vvs, br) | 355 (vvs, sbr) | 352 (vvs, sbr) |
| | $\delta_{MnCN}$ | $(F_{1u})$ | 468 (st) | 471 (st) | 467 (st) |
| $K_4[Mn(CN)_6] \cdot 3H_2O$ | $\nu_{CN}$ | $(F_{1u})$ | 2032 (st) | 2037 (st) | 2032 (st) |
| | $\nu_{MnC}$ | $(F_{1u})$ | 376 (vvs, br) | 381 (vvs, br) | 376 (vvs, br) |
| | $\delta_{MnCN}$ | $(F_{1u})$ | 515 (m) | | |
| $K_3[Fe(CN)_6]$ | $\nu_{CN}$ | $(F_{1u})$ | 2110 (vvs) | 2114 (st), 2061 (sh), 2045 (st, br) | 2110 (st), 2037 (st, br) |
| | $\nu_{FeC}$ | $(F_{1u})$ | 386 (vvs, br) | 394 (vvs, br) | 386 (vvs, br) |
| | $\delta_{FeCN}$ | $(F_{1u})$ | 507 (st) | 509 (st), 585 (st) | 507 (st), 581 (st) |

| Compound | | | | |
|---|---|---|---|---|
| [N(CH$_3$)$_4$]$_3$[Fe(CN)$_6$] | $\nu_{CN}$ ($F_{1u}$) | 2110 (vvs) | 2114 (st), 2049 (st) | 2105 (st), 2033 (st) |
| K$_4$[Fe(CN)$_6$]·3H$_2$O | $\nu_{CN}$ ($F_{1u}$) | 2037 (vvs) | 2055 (vvs, br) | 2037 (vvs), 2024 (sh) |
|  | $\nu_{FeC}$ ($F_{1u}$) | 413 (vvs), 406 (sh) | 415 (vvs) | 413 (vvs), 406 (sh) |
|  | $\delta_{FeCN}$ ($F_{1u}$) | 580 (vvs) | 585 (vvs) | 580 (vvs) |
| [N(CH$_3$)$_4$]$_4$[Fe(CN)$_6$]·$n$H$_2$O | $\nu_{CN}$ ($F_{1u}$) | 2033 (vvs) | 2050 (vvs) | 2035 (vvs), 2020 (sh) |
| K$_3$[Co(CN)$_6$] | $\nu_{CN}$ ($F_{1u}$) | 2124 (st), 2099 (vs) | 2129 (st), 2099 (sh) | 2122 (st), 2099 (vs) |
|  | $\nu_{CoC}$ ($F_{1u}$) | 410 (st) | 415 (st) | 411 (st) |
|  | $\delta_{CoCN}$ ($F_{1u}$) | 562 (m) | 563 (m) | 562 (m) |
| K$_2$[Ni(CN)$_4$] | $\nu_{CN}$ ($E_u$) | 2115 (st) | 2128/2120 (st) | 2115 (st) |
|  | $\nu_{NiC}$ ($E_u$) | 408 (vvs, br) | 412 (vvs, br) | 407 (vvs, br) |
|  | $\delta_{NiCN}$ ($E_u$) | 524 (vs) | 526 (vs) | 525 (vs) |
| K$_2$[Zn(CN)$_4$] | $\nu_{CN}$ ($F_2$) | 2144 (st), 2101 (vs) | 2160 (m), 2151 (st) | 2144 (st) |
|  | $\nu_{ZnC}$ ($F_2$) | 350 (vvs, br) | 348 (vvs, br) | 350 (vvs, br) |
|  | $\delta_{ZnCN}$ ($F_2$) | 312 (m–st) | 313 (m–st) | 311 (m–st) |

[a] Abbreviations: br = broad; m = medium; sh = shoulder; st = strong; vs = very strong; vvs = very very strong.

### Table XVII

*Pressure Effects on Skeleton Modes of Metal Sandwich Compounds*

| Compound | Ring tilt (cm$^{-1}$) | | $\nu$ (M-ring) (cm$^{-1}$) | |
|---|---|---|---|---|
| | Ambient pressure | 35 kbar | Ambient pressure | 35 kbar |
| Fe(Cp)$_2$ | 491 | 494 | 461 | 474 |
| Ru(Cp)$_2$ | 447 | 476 | 381 | 390 |
| Mn(Cp)$_2$ | 432 | 437 | 409 | 420 |
| Cr(C$_6$H$_6$)$_2$ | 487 | 491 | 453 | 475 |

nal $\nu$CoX. The $\nu$CoN shifts were smaller, Table XVIII lists the pressure effects obtained for these complexes.

(h) *Ligand Vibrations* Bayer and Ferraro (1969) studied pressure effects on the vibrations of pyrazine up to 72 kbar. Since pyrazine tends to ligate transition elements, the pressure effects were compared with those obtained for metal complexes of pyrazine. The following observations were made.

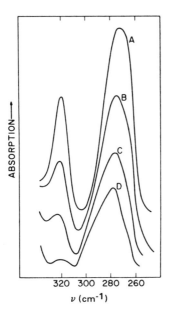

**Fig. 8.** Frequency of $\nu_{GeCl_{asym}}$ and $\nu_{GeCl_{sym}}$ in [$\phi_4$As][GeCl$_3$] as a function of pressure in a DAC. Ge–Cl stretching bands for [(C$_6$H$_5$)$_4$As]–[GeCl$_3$], a trigonal-pyramidal structure. Curves are displaced vertically: A, atmospheric pressure (320, 272 cm$^{-1}$); B, 4600 atm (322, 275 cm$^{-1}$; C, 10,800 atm (324, 278 cm$^{-1}$); D, 17,000 atm (322, 278 cm$^{-1}$). [Reprinted with permission from Postmus *et al.* (1967). Copyright 1967 American Chemical Society.]

(a) Ligand vibration showed blue shifts and were less pressure sensitive than those in the complexes.
(b) Splitting occurs for some bands with pressure.
(c) The pyrazine spectrum under pressure resembled that of the spectrum of the corresponding metal complex.

*Table XVIII*

*Pressure Effects on Low-Frequency Infrared Vibration-Bridging, Metal–Ligand, and Ligand Motions*

| Polymeric compound | Assignment | Ambient pressure | Nonambient pressure[a] | Shift (cm$^{-1}$) |
|---|---|---|---|---|
| Co(py)$_2$Cl$_2$ | $\nu_{CoN}$ | 243 | 260 | 17 |
| | | 235 | Disappears | — |
| | L[b] | 225 | Disappears | — |
| | $\nu_b$CoCl | 174 | 220 | 46 |
| | | 186 | Disappears | — |
| Co(4-Br(py))$_2$Br$_2$ | $\nu_b$CoBr | 143 | 173 | 31 |
| | | 128 | 142 | 14 |
| | L | 195 | 199 | 4 |
| | $\nu_{CoN}$ | 228 | 230 | 2 |
| | L | 290 | 292 | 2 |
| Co(4-Br(py))$_2$Cl$_2$ | $\nu_b$CoCl | 155 | Disappears | — |
| | | 178 | 200 | 22[c] |
| | L | 207 | 208 (sh) | 1 |
| | $\nu_{CoN}$ | 232 | 240, 247 | 11 |
| | L | 290 | 293 | 3 |
| Co(4-Cl(py))$_2$Cl$_2$ | $\nu_b$CoCl | 165 | 190 | ~35 |
| | | 185 | Shifts into 217 band | ~32[c] |
| | L | 217 | 217 | 0 |
| | $\nu_{CoN}$ | 241 | 259 | 18 |
| | L | 322 | 326 | 4 |
| Co(4-Cl(py))$_2$Br$_2$ | $\nu_b$CoBr | 118 (sh)[d] | 146 | 14 |
| | | 138 | 175 | 37 |
| | L | 207 | 209, 223 (sh) | 2 |
| | $\nu_{CoN}$ | 238 | 245 | 7 |
| | L | 322 | 323 | 1 |

[a] All pressures are 28 kbars, except for Co(py)$_2$Cl$_2$, where 36-kbar pressures were used.
[b] L = ligand.
[c] Estimated.
[d] sh = shoulder.

Other organic substances that can be considered as ligating compounds are discussed in Chapter 6.

## REFERENCES

Adams, D. M., and Payne, S. J. (1975). *J. Chem. Soc. Dalton Trans.*, 215.
Adams, D. M., and Payne, S. J. (1976). *Inorg. Chim. Acta* **19,** L49.
Adams, D. M., Payne, S. J., and Martin, K. (1973). *Appl. Spectrosc.* **27,** 377.
Adams, D. M., Findley, J. D., Coles, M. C., and Payne, S. J. (1976). *J. Chem. Soc. Dalton Trans.*, 371.
Adams, D. M., Berg, R. W., and Williams, A. D. (1981a). *J. Chem. Phys.* **74,** 2800.
Adams, D. M., Hatton, P. D., Shaw, A. C., and Tan, T-K. (1981b). *J. Chem. Soc. Chem. Commun.* 226.
Adams, D. M., Gerrard, M. E., and Hatton, P. D. (1981c). *Solid State Commun.* **39,** 229.
Adams, D. M., Long, G. J., and Williams, A. D. (1982). *Inorg. Chem.* **21,** 1049.
Allkins, J. R., Obremski, R. J., Brown, C. W., and Lippincott, E. R. (1969). *Inorg. Chem.* **8,** 1450.
Anderson, D. N., and Willett, R. G. (1974). *Inorg. Chim. Acta* **8,** 167.
Bader, R. F. W. (1960). *Mol. Phys.* **3,** 137.
Bader, R. F. W. (1962). *Can. J. Chem.* **40,** 1164.
Bargeron, C. B., and Drickamer, H. G. (1971). *J. Chem. Phys.* **55,** 3471.
Bargeron, C. B., Avinor, M., and Drickamer, H. G. (1971). *Inorg. Chem.* **10,** 1338.
Basile, L. J., Ferraro, J. R., Choca, M., and Nakamoto, K. (1974). *Inorg. Chem.* **13,** 496.
Bayer, R., and Ferraro, J. R. (1969). *Inorg. Chem.* **8,** 1654.
Blight, D. G., and Kepert, D. L. (1968). *Theor. Chim. Acta* **11,** 51.
Breitinger, D., and Kristen, U. (1977). *J. Solid State Chem.* **22,** 233.
Brock, C. P., Collman, J. P., Dolcetti, G., Farnham, P. H., Ibers, J. A., Lester, J. E., and Reed, C. A. (1973). *Inorg. Chem.* **12,** 1304.
Browning, M. C., Mellor, J. R., Morgan, D. J., Pratt, S. A. J., Sutton, L. E., and Venanzi, L. M. (1962). *J. Chem. Soc.*, 693.
Burbank, R. D., and Bartlett, N. (1968). *Chem. Commun.*, 645.
Butcher, R. J., Ferraro, J. R., and Sinn, E. (1976a). *Inorg. Chem.* **15,** 2077.
Butcher, R. J., Ferraro, J. R., and Sinn, E. (1976b). *J. Chem. Soc. Chem. Commun.*, 910.
Claassen, H. H., Gasner, E. L., and Selig, H. (1968). *J. Chem. Phys.* **49,** 1803.
Dehnicke, G., Dehnicke, K., Ahsbahs, H., and Hellner, E. (1974). *Ber. Bunsenges. Phys. Chem.* **78,** 1010.
Drickamer, H. G. (1965). *Solid State Phys.* **17,** 1.
Drickamer, H. G., and Frank, C. W. (1973). "High Pressure Chemistry and Physics of Solids." Chapman and Hall, London.
Earnshaw, A., King, E. A., and Larkworthy, L. F. (1969). *J. Chem. Soc. A,* 2459.
Eisenberg, R., and Gray, H. B. (1967). *Inorg. Chem.* **6,** 1844.
Eisenberg, R., and Ibers, J. A. (1965). *J. Am. Chem. Soc.* **87,** 3776.

# References

Feltham, R. D., Enemark, J. H., Ferraro, J. R., and Basile, L. J. (1979). Unpublished data.
Ferraro, J. R. (1970). *J. Chem. Phys.* **53,** 117.
Ferraro, J. R. (1976). *J. Coord. Chem.* **5,** 101.
Ferraro, J. R. (1975). Argonne National Lab., Argonne, Illinois, unpublished data.
Ferraro, J. R., and Fabbrizzi, L. (1978). *Inorg. Chim. Acta* **26,** 615.
Ferraro, J. R., and Long, G. J. (1975). *Accts. Chem. Res.* **8,** 171.
Ferraro, J. R., and Nakamoto, K. (1972). *Inorg. Chem.* **11,** 2290.
Ferraro, J. R., and Sherren, A. (1978). *Inorg. Chem.* **17,** 2498.
Ferraro, J. R., and Takemoto, J. (1974). *Appl. Spectrosc.* **28,** 66.
Ferraro, J. R., Meek, D. W., Siwiec, E. C., and Quattrochi, A. (1971). *J. Am. Chem. Soc.* **93,** 3862.
Ferraro, J. R., Nakamoto, K., Wang, J. T., and Lauer, L. (1973). *J. Chem. Soc. Chem. Commun.* 266.
Ferraro, J. R., Basile, L. J., and Sacconi, L. (1979). *Inorg. Chim. Acta* **35,** 6317.
Fisher, D. C., and Drickamer, H. G. (1971). *J. Chem. Phys.* **54,** 4825.
Fleischer, E. B., Gebala, A. E., Swift, D. R., and Tasker, P. A. (1972). *Inorg. Chem.* **11,** 2775.
Frank, C. W., and Drickamer, H. G. (1972). *J. Chem. Phys.* **56,** 3551.
Fung, S. C., and Drickamer, H. G. (1969a). *J. Chem. Phys.* **51,** 4353.
Fung, S. C., and Drickamer, H. G. (1969b). *Proc. Natl. Acad. Sci. U.S.A.* **62,** 38.
Haller, K. J., Johnson, P. E., Feltham, R. D., Enemark, J. H., Ferraro, J. R., and Basile, L. J. (1979). *Inorg. Chim. Acta* **33,** 119.
Hartman, K. O., and Miller, F. A. (1968). *Spectrochim. Acta* **24,** 669.
Hayes, R. G. (1966). *J. Chem. Phys.* **44,** 2210.
Hellner, E., Ahsbahs, H., Dehnicke, G., and Dehnicke, K. (1973). *Ber. Bunsenges. Phys. Chem.* **77,** 277.
Hellner, E., Ahsbahs, H., Dehnicke, G., and Dehnicke, K. (1974). *Naturwiss.* **61,** 502.
Hoard, J. L., and Silverton, J. V. (1963). *Inorg. Chem.* **2,** 235.
Hoard, J. L., Hamar, T. A., and Glick, M. D. (1968). *J. Am. Chem. Soc.* **90,** 3172.
Holmes, R. R. (1972). *Accts. Chem. Res.* **5,** 296.
Kilbourn, B. T., and Powell, H. M. (1970). *J. Chem. Soc. A* 1668.
Kilbourn, B. T., Powell, H. M., and Darbyshire, J. A. C. (1963). *Proc. Chem. Soc., London* 207.
Klanberg, F., Eaton, D. R., Guggenberger, L. J., and Muetterties, E. L. (1967). *Inorg. Chem.* **6,** 1271.
Lippard, S. J. (1966). *Prog. Inorg. Chem.* **8,** 109.
Long, G. J., and Coffen, D. L. (1974). *Inorg. Chem.* **13,** 270.
Long, G. J., and Ferraro, J. R. (1973). *J. Chem. Soc. Chem. Commun.* 719.
Long, G. J., and Ferraro, J. R. (1974). *Inorg. Nucl. Chem. Lett.* **10,** 393.
Long, G. J., and Schlemper, E. O. (1974). *Inorg. Chem.* **13,** 279.
Longuet-Higgins, H. C. (1956). *Proc. R. Soc., London, Ser. A* **235,** 537.
McGarvey, B. R. (1966). *Inorg. Chem.* **5,** 476.
Malatesta, L., Fermi, M., and Valesti, V. (1964). *Gazz. Chim. Ital.* **94,** 1278.
Muetterties, E. L. (1970a). *Acc. Chem. Res.* **3,** 266.
Muetterties, E. L. (1970b). *Rec. Chem. Prog.* **31,** 51.

Muetterties, E. L. (1973). *Inorg. Chem.* **12,** 1963.
Muetterties, E. L., and Schunn, R. A. (1966). *Q. Rev. Chem. Soc. (London)* **20,** 245.
Muetterties, E. L., and Wright, C. M. (1967). *Q. Rev. Chem. Soc. (London)* **21,** 109.
Nakamoto, K., Udovich, C., Ferraro, J. R., and Quattrochi, A. (1970). *Appl. Spectrosc.* **24,** 606.
Parish, R. V. (1966). *Coord. Chem. Rev.* **1,** 439.
Parish, R. V., and Perkins, P. G. (1967). *J. Chem. Soc.,* 345.
Pearson, R. G. (1969a). *J. Am. Chem. Soc.* **91,** 4947.
Pearson, R. G. (1969b). *J. Am. Chem. Soc.* **91,** 1252.
Pearson, R. G. (1970). *J. Chem. Phys.* **53,** 2986.
Pearson, R. G. (1971a). *Pure Appl. Chem.* **27,** 145.
Pearson, R. G. (1971b). *Accts. Chem. Res.* **4,** 152.
Pearson, R. G. (1972). *J. Am. Chem. Soc.* **94,** 8287.
Pearson, R. G. (1976). "Symmetry Rules for Chemical Reactions: Orbital Topology and Elementary Processes." Wiley (Interscience), New York.
Postmus, C., Nakamoto, K., and Ferraro, J. R. (1967). *Inorg. Chem.* **6,** 2194.
Raymond, K. W., Corfield, P. W. R., and Ibers, J. A. (1968). *Inorg. Chem.* **7,** 1362.
Rossman, G. R., Tsay, F. D., and Gray, H. B. (1973). *Inorg. Chem.* **12,** 824.
Sacconi, L., and Ferraro, J. R. (1974). *Inorg. Chim. Acta* **9,** 49.
Sinn, E. (1974). *Coord. Chem. Rev.* **12,** 185.
Smith, A. E., Schrauzer, G. N., Hayweg, V. P., Heinrich, W. (1965). *J. Am. Chem. Soc.* **87,** 5798.
Vaughan, R. W., and Drickamer, H. G. (1967). *J. Chem. Phys.* **47,** 468.
Wang, P. J., and Drickamer, H. G. (1973). *J. Chem. Phys.* **59,** 559.
Willett, R. D., Ferraro, J. R., and Choca, M. (1974). *Inorg. Chem.* **13,** 2919.
Wong, P. T. T. (1979). *J. Chem. Phys.* **70,** 456.
Wood, J. C. (1972). *Prog. Inorg. Chem.* **16,** 227.

## BIBLIOGRAPHY

*Pressure Studies on Mercurimethanes*

Breitinger, D. K., Morrell, W., and Grabetz, K. (1979). *Z. Naturforsch.* **346,** 390.
Breitinger, D. K., Geibel, K., Kress, W. and Sendelbeck, R. (1980). *J. Organomet. Chem.* **191,** 7.

*Other Coordination Compounds*

Armstrong, R. L. (1971). *J. Chem. Phys.* **54,** 813.
Hara, Y., and Minomura, S. (1974). *J. Chem. Phys.* **61,** 5339.

CHAPTER 6
# ORGANIC AND BIOLOGICAL COMPOUNDS

In comparison with the attention that has been devoted to solid inorganic and coordination compounds, the examination of solid organic and biological compounds at high pressures has lagged behind. A great deal of the research has been done in the organic fingerprint region of the infrared, and one deals primarily with intramolecular vibrations (internal modes). However, many of the same phenomena occurring with inorganic or coordination compounds can also occur with organic or biological molecules. For example, frequency increases with pressure (similar to those modes occurring at lower temperatures in inorganic or coordination compounds, although those are of larger magnitude) can occur. Changes in shapes of vibrational bands are also possible (peak heights and integrated intensities may be altered), and splitting of bands can occur. When two bands have differing pressure coefficients, modification of Fermi resonance may occur. This latter phenomenon was first observed by Sherman and Smulovitch (1970). Phase transitions occurring with pressure are possible, with a consequence of a differing vibrational spectrum due to changes in selection rules. Conformational changes may also occur with pressure. Molecular lattice modes appear to be more pressure sensitive than ionic lattice modes.

## 1. ORGANIC MOLECULES

The amount of research conducted with the DAC and dealing with solid organic molecules involving vibrational spectroscopy is rather limited. We shall discuss pressure effects on lubricants in Chapter 7. Certain review articles on pressure effects on organic compounds are listed at the end of this chapter. Many of these deal with solution studies and involve neither vibrational spectroscopy nor the DAC.

### a. Iodoform

Bockelmann and Sherman (1980) observed the lattice mode shifts with pressure (see Fig. 1). These molecular lattice modes demonstrated significant pressure dependencies. For example, the 32-cm$^{-1}$ vibration showed a $d\nu/dp = 1.6$ cm$^{-1}$ kbar$^{-1}$ and the 37-cm$^{-1}$ vibration a $d\nu/dp = 1.5$ cm$^{-1}$ kbar$^{-1}$. Medina *et al.* (1982) have made single-crystal Raman measurements of iodoform at 10–300 K temperatures and up to 14-kbar pressure. Grüneisen parameters of lattice modes were found to show higher values than those for internal modes.

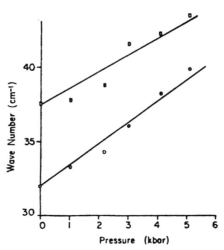

*Fig. 1.* The shifts in wave number of the lowest Raman bands of iodoform with pressure. (□, shifts of 37-cm$^{-1}$ band; ○, shifts in 32-cm$^{-1}$ band.) [From Bockelmann and Sherman (1980).]

## b. $CCl_4$, $CBr_4$

Adams and Sharma (1976) have made a Raman study of the solid phase of $CCl_4$ and $CBr_4$ under pressure. Four solid phases have been identified for $CCl_4$, and the phase diagram ($T$ versus $p$) for $CCl_4$ is depicted in Fig. 2. Figure 3 shows the Raman spectra of the various phases.

Three phases for $CBr_4$ were identified. Phase III was not characterized because it photodecomposed rapidly. Figure 4 shows the phase diagram, and Fig. 5 illustrates the Raman spectra of the three phases.

From the data, inferences for a $SnI_4$ structure for $CCl_4(IV)$ and $CBr_4(IV)$ were drawn.

## c. Methanol and Ethanol

Mammone *et al.* (1980) examined the Raman spectra of methanol and ethanol under pressures of up to 100 kbar. Methanol crystallized or formed a glass. When it crystallized it did so at $35.1 \pm 1.0$ kbar. The crystal phase was similar to the low-pressure $\alpha$ phase. Ethanol crystallized at $17.8 \pm 1.0$ kbar, and the crystal phase was interpreted as was the low-temperature monoclinic phase. The hydroxyl stretching frequencies demonstrated large negative shifts with pressure, indicative of the occurrence of stronger hydrogen bonding.

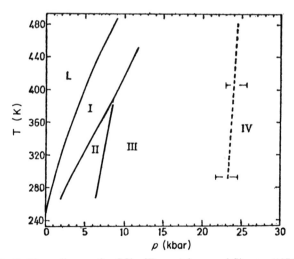

*Fig. 2.* Phase diagram for $CCl_4$. [From Adams and Sharma (1976).]

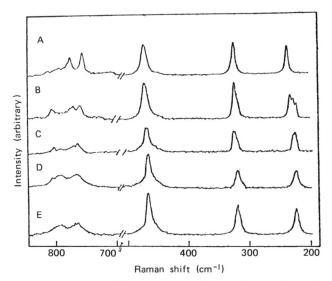

**Fig. 3.** Raman spectra of CCl$_4$ polymorphs. See text for conditions of preparation. All the spectra were obtained with 300-mW 514.5-nm radiation at the sample. Spectral slit width was 1.0–1.2 cm$^{-1}$ for wave numbers of 200–500 cm$^{-1}$ and 2.5 cm$^{-1}$ for wave numbers of 500–800 cm$^{-1}$. Phases: (A) IV, 24.3 kbar (293 K); (B) III, 12.8 kbar; (C) II, 6.25 kbar; (D) I + liquid, 1.12 kbar; and (E) liquid, 0.66 kbar. [From Adams and Sharma (1976).]

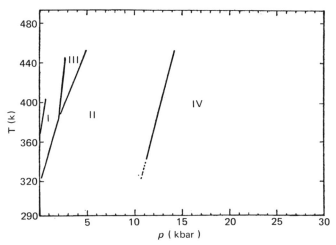

**Fig. 4.** Phase diagram for CBr$_4$. [From Adams and Sharma (1976).]

# 1. Organic Molecules

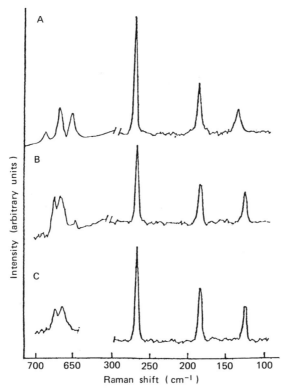

*Fig. 5.* Raman spectra of $CBr_4$ polymorphs. All the spectra were obtained with 100-mW 647.1-nm radiation at the sample. Spectral slit width 1.2 cm$^{-1}$ for wave numbers of 100–300 cm$^{-1}$ and 2.5 cm$^{-1}$ for wave numbers of 600–700 cm$^{-1}$. Phases: (A) IV, 20.4 kbar (293 K); (B) II, 0.001 kbar (293 K); and (C) I, 0.001 kbar (322.4 K). [From Adams and Sharma (1976).]

## d. Benzene

Ellenson and Nicol (1974) characterized the solid phase of benzene (II) obtained at 17 kbar in a Drickamer-type cell. Raman spectroscopy was the diagnostic probe. Adams and Appleby (1977) examined the three solid phases of benzene by using the DAC and MIR, FIR, and Raman spectroscopy. Phase III was produced at ~40-kbar pressure and 100°C. The latter phase transition was found to occur without a change in space group or cell content but with a volume change ($\Delta V/V \simeq 0.15$), indicative of a first-order transition

**Table I**

*Crystallographic Data on the Three Solid Phases of Benzene*

| Condensed phase of $C_6H_6$ | Space group | $Z'$ |
|---|---|---|
| I | $D_{2h}^{15}$ (Pbca) | 4 |
| II | $C_{2h}^{5}$ (P2$_1$/c) | 2 |
| III | Probably like phase II | |

(see Adams and Appleby, 1977). Table I lists crystallographic data for benzene I and II, and Table II shows the selection rules for phases I and II. Figure 6 shows the Raman data, Fig. 7 the FIR data in the lattice mode region, and Fig. 8 the MIR data for the three phases. Table III summarizes the infrared data and assignments for various phases in the MIR region.

### e. p-Dihalogenbenzenes and Related Molecules

The $p$-dihalogenbenzenes have been investigated under pressure by Adams and Ekejiuba (1981), who used Raman spectroscopy. For $p$-dichlorobenzene an $\alpha$- to $\gamma$-phase transition was observed at ~1.5 kbar. A second-order transition to a $\delta$ phase occurred at ~9 kbar. Figure 9 shows the Raman spectra of $p$-$C_6H_4Cl_2$ at several pressures in the lattice region, and Fig. 10 demonstrates the frequency varia-

**Table II**

*Selection Rules for Benzene I and Benzene II Lattice Modes*

Benzene I [a,b] $3\underline{A_u} + 3\underline{B_{1u}} + 3\underline{B_{2u}} + 3\underline{B_{3u}} + 3A_g + 3B_{1g} + 3B_{2g} + 3B_{3g}$

($B_{1u} + B_{2u} + B_{3u}$ are acoustic modes)

Benzene II [a] $3A_u + 3B_u + 3A_g + 3B_g$ ($A_u + 2B_u$ are acoustic modes)

Benzene I has $3nZ' = 3.12.4 = 144$ total modes and $(3n - 6)Z' = (3.12 - 6)4 = 120$ are internal modes; therefore there are 24 external modes

Benzene II has $3nZ' = 3.12.2 = 72$ total modes and $(3n - 6)Z' = (3.12 - 6)2 = 60$ are internal modes; therefore there are 12 external modes

[a] Considering a $C_i$ site for benzene.
[b] Underlined species unallowed in Raman or IR.

## 1. Organic Molecules

### Table III

*IR Wave Numbers (Cm$^{-1}$) for the Solid Phases of Benzene at Ambient Temperature*

| $C_6H_6I$, 1.6 kbar | $C_6H_6II$, 20.3 kbar | $C_6H_5III$, 34.4 kbar | Assignment |
| --- | --- | --- | --- |
| 1830 | — | — | $\nu_{10} + \nu_{17}$ |
| 1755 | — | — | $\nu_6 + \nu_{15}$ |
| 1715 | — | — | $\nu_4 + \nu_{12}$ |
| 1680 | 1680 | 1690 | $\nu_4 + \nu_{17}, \nu_1 + \nu_{11}$ |
| 1620 | 1620 | 1628 | $\nu_6 + \nu_{12}$ |
| 1547 | 1550 | 1567 | $\nu_{10} + \nu_{11}$ |
| 1478 | 1480 | 1486 | $\nu_{19}$ |
| 1402 | 1411 | 1442 | $\nu_5 + \nu_{16}$ |
| 1310 | 1325 | 1340 | $\nu_{14}$ |
| 1260 | 1255 [a] | 1276 | $\nu_{10} + \nu_{16}$ |
| 1210 | — | — | — |
| 1180 | — | — | — |
| 1149<br>1144 | {1170 | 1186 | $\nu_{15}$ |
| 1035 | 1045 | { 1065<br>{ 1047 | $\nu_{18}$ |
| 1011 | 1016 | 1019 | $\nu_{12}$ |
| 981<br>975 | ⎡ 991<br>⎢ 985<br>⎨ 972<br>⎣ 967 | ⎡ 1002<br>⎢ 998<br>⎨ 977<br>⎣ 970 | $\nu_{17}$ |
| 770 | —[b] | — | $\nu_{11} + \nu_L$ |
| 680 | ~685 | 695 | $\nu_{11}$ |

[a] Further unresolved components to high frequency.
[b] Broad region of absorption.

tion with pressure. An equivalent α- to γ-phase transition was not detected for p-$C_6H_4Br_2$, but a discontinuity at ~5 kbar was observed when $\Delta\nu$ was plotted versus pressure (see Fig. 11). In Fig. 12 the Raman spectra for p-$C_6H_4Br_2$ under pressure is illustrated. Earlier work on 12 *para*-substituted dihalogenbenzenes and related compounds of the type

$$X-\bigcirc-Y$$

where X and Y are Cl, Br, or I and $NO_2$ or $CH_3$, respectively, were reported by Hamann (1978). He used infrared spectroscopy to deter-

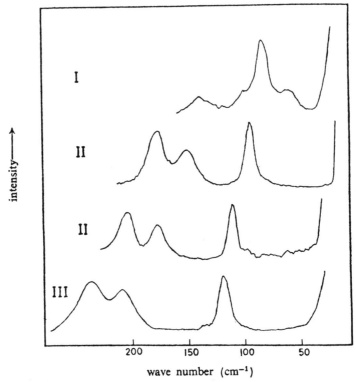

*Fig. 6.* Raman spectra of solid phases of benzene in the lattice mode region: Benzene I at 295 K and 5.3 kbar; spectral slit width 1.7 cm$^{-1}$, 50-mW, 514.5-nm radiation at the sample. Benzene II (upper curve) at 295 K and 20.3 kbar; spectral slit width 2.4 cm$^{-1}$, 250 mW at the sample. Benzene II (lower curve) at 136 K and 32.8 kbar; spectral slit width 1.9 cm$^{-1}$, 400 mW, 514.5 nm at the sample. Benzene III at 296 K and 34.4 kbar; spectral slit width 2.4 cm$^{-1}$, 250 mW, 514.5 nm at the sample. [From Adams and Appleby (1977).]

mine the nature of phase transitions occurring up to pressures of 40 kbar.

Hamann (1976) also examined seven halogen-substituted and one methyl-substituted benzene compound in a DAC with pressure of up to ~50 kbar. The compounds studied were *p*-dichlorobenzene; 1,3,5-trichlorobenzene; 1,2,4,5-tetrachlorobenzene; hexachlorobenzene; hexabromobenzene; *p*-dibromobenzene; 1,3,5-tribromobenzene; 1,2,4,5-tetrabromobenzene; and hexamethylbenzene. Infrared spectroscopy provided the means of detecting changes. All com-

# 1. Organic Molecules

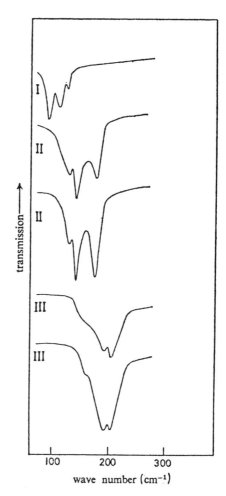

*Fig. 7.* FIR spectra of solid phases of benzene: Benzene I at 295 K and 4.1 kbar. Benzene II (upper curve) at 295 K and 20.3 kbar and (lower curve) at 188 K and 20.3 kbar. Benzene III (upper curve) at 295 K and 34.4 kbar and (lower curve) at 188 K and 34.4 kbar. [From Adams and Appleby (1977).]

pounds, with the exception of $C_6(CH_3)_6$, showed signs of pressure-induced phase transitions. Hamann formulated from this work certain generalizations about what occurs with an increase in pressure:

(a) Original IR bands shift to higher frequency with pressure.
(b) Peak heights of bands decrease with increase in pressure.
(c) Integrated intensities of bands are greatly altered. All decrease except for the C–H stretching vibrations, which become more intense.

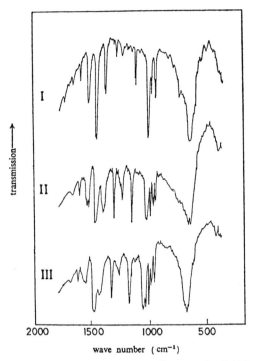

***Fig. 8.*** MIR spectra at ambient temperature of benzene I at 1.6 kbar, benzene II at 20.3 kbar, and benzene III at 34.4 kbar. Spectral slit width is 3.7 cm$^{-1}$ throughout. Spectra for benzene II and III were obtained by using the same samples as were used for the Raman spectra of Fig. 6 [benzene II (upper curve) and benzene III] and the FIR spectra of Fig. 7 [benzene II (upper curve) and benzene III (upper curve)]. The rising background at the high-frequency end is due to two-phonon absorption in the diamond anvils. [From Adams and Appleby (1977).]

(d) A ring-bending vibration, which is inactive in the IR for 1,3,5-trichlorobenzene and 1,3,5-tribromobenzene in the ambient phase, becomes active with high pressure. However, the band disappears with pressure, in 1,2,4,5-tetrachlorobenzene and the bromine analog, where it is normally IR active in the ambient phase.

### f. Naphthalene, Octafluoronaphthalene

Naphthalene crystallizes in a $C_{2h}^5$ space group with $Z' = 2$ with $C_i$ sites. Selection rules are illustrated in Table IV. Wu and Jura (1973)

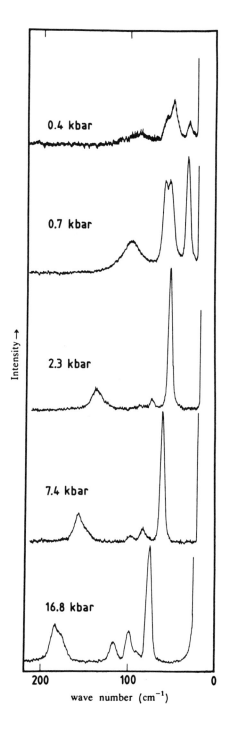

**Fig. 9.** Raman spectra of $p$-$C_6H_4Cl_2$ at various pressures. Spectral slit width is 1.5 cm$^{-1}$. [From Adams and Ekejiuba (1981).]

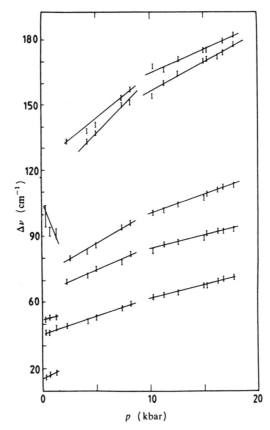

**Fig. 10.** Variation of mode frequencies with pressure for $p$-$C_6H_4Cl_2$. [From Adams and Ekejiuba (1981).]

**Table IV**

*Selection Rules for Naphthalene*

| | |
|---|---|
| $T'_{internal}$ | $24A_g + 24A_u + 24B_g + 24B_u$ |
| $T'_{external}$ | $3A_g + 3B_g + 3A_u + 3B_u$ |
| | ($2A_u + B_u$ = acoustic modes) |

# 1. Organic Molecules

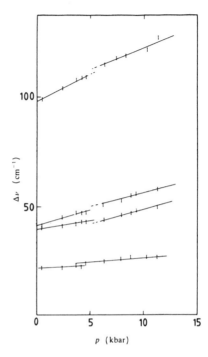

*Fig. 11.* Variation of mode frequencies with pressure for $p$-$C_6H_4Br_2$. [From Adams and Ekejiuba (1981).]

examined naphthalene under pressure in the MIR, and Dows *et al.* (1973) studied the pressure dependency of the Raman-active lattice modes to 10 kbar. Figure 13 shows the pressure and volume dependence of the lattice modes. All six external modes showed a frequency increase with pressure. Bandwidths did not change with pressure, and no new phases were observed in the pressure range studied. The internal modes were less pressure sensitive. Figure 14 shows the infrared spectra of naphthalene in the 1000-$cm^{-1}$ region.

Adams *et al.* (1980) obtained Raman and MIR spectra of octafluoronaphthalene ($C_{10}F_8$) under hydrostatic pressures up to 17 kbar in a DAC. A phase change was observed to occur at 0.8 kbar. Figure 15 shows the Raman spectra of $C_{10}F_8$ under pressure. Pressure dependencies of Raman-active lattice modes are shown in Fig. 16, and pressure dependencies of internal modes (IR and Raman) are illustrated in Fig. 17.

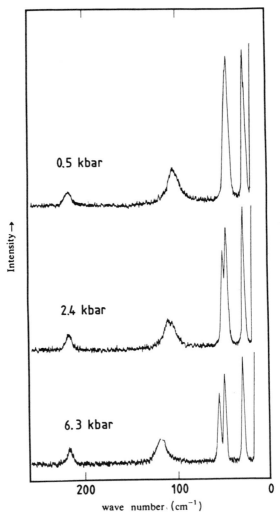

**Fig. 12.** Raman spectra of *p*-C$_6$H$_4$Br$_2$ at various pressures. Spectral slit width is 1 cm$^{-1}$. [From Adams and Ekejiuba (1981).]

### g. Methane Ice

Methane ice has a space group $O_h^5$ (*Fm3m*), where the primitive unit cell involves one molecule (see Hazen *et al.*, 1979–1980). Space groups previously reported for solid CH$_4$ weer $T_d^2$ by Anderson and Walmsley (1965). The low-temperature phase is designated as phase

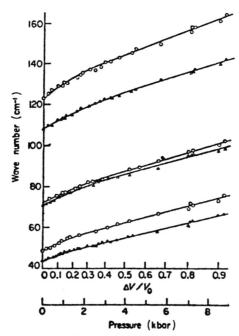

*Fig. 13.* The pressure and volume dependence of the Raman-active lattice vibrations of naphthalene. ○, $B_g$ mode wave numbers; ▲, $A_g$ mode numbers. [From Dows et al. (1973). Copyright North-Holland Publishing Company, Amsterdam, 1973.]

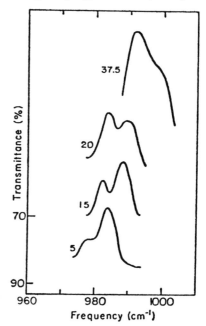

*Fig. 14.* The shift in the wave number of the infrared bands of naphthalene at ~1000 cm$^{-1}$ with pressure given in kilobar. [From Wu and Jura (1973). Reprinted with permission from *Spectrochimica Acta, Part A*, Vol. 30. Copyright 1973, Pergamon Press, Ltd.]

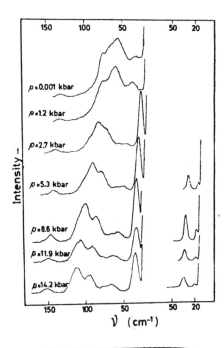

**Fig. 15.** Raman spectra of $C_{10}F_8$ at various pressures in a DAC in the region of the lattice modes. Spectral slit width is 3 cm$^{-1}$. Inset shows the two bands characteristic of phase II; spectral slit width is 2 cm$^{-1}$. [From Adams et al. (1980). Reprinted with permission from *Journal of Physics and Chemistry of Solids*, Vol. 41. Copyright 1980, Pergamon Press, Ltd.]

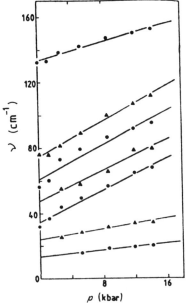

**Fig. 16.** Pressure dependencies of the Raman-active lattice-mode phonons of $C_{10}F_8$. [From Adams et al. (1980). Reprinted with permission from *Journal of Physics and Chemistry of Solids*, Vol. 41. Copyright 1980, Pergamon Press, Ltd.]

## 1. Organic Molecules

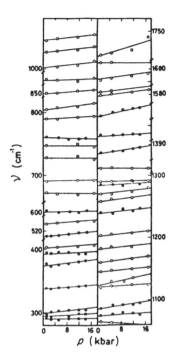

*Fig. 17.* Pressure dependencies of the internal mode frequencies in $C_{10}F_8$. (○) IR, (●) Raman. [From Adams *et al.* (1980). Reprinted with permission from *Journal of Physics and Chemistry of Solids,* Vol. 41. Copyright 1980, Pergamon Press, Ltd.]

I. Other phases are II through IV. Constantino and Daniels (1975) have studied the low-temperature phase to a pressure limit of 10 kbar. Raman studies under pressure (to 187 kbar) of several of the other phases were made by Sharma *et al.* (1979–1980). Table V tabulates the results. The transition from phase I to phase V was observed to occur with an abrupt decrease of the half-bandwidth of $\nu_3$ from 64 cm$^{-1}$ in phase I to 32 cm$^{-1}$ in phase V. Additionally, the $\nu_2$ band decreases in frequency from 1551 cm$^{-1}$ (phase I) to 1536 cm$^{-1}$ (phase V). On this basis, it was suggested by Sharma *et al.* (1979–1980) that the transition from phase I to phase V involves a first-order transition. With higher pressures, all Raman bands shift to higher frequency. The $\nu_3$ band in phase V is the most pressure sensitive and demonstrates a $d\nu/dp$ of 1.3 cm$^{-1}$ kbar$^{-1}$; next is the $\nu_1$ mode at 0.95 cm$^{-1}$ kbar$^{-1}$; $\nu_2$ is the least pressure sensitive, with $d\nu/dp = 0.19$ cm$^{-1}$ kbar$^{-1}$. At 118 kbar a weak shoulder appears at 2991 cm$^{-1}$ on the $\nu_1$ band. At 187 kbar, two weak shoulders appear on the $\nu_3$ band as well as the $\nu_1$ band (see Table V). These results are indica-

## Table V
### Raman Frequencies of Methane at Various Pressures and Room Temperature

| Band | 0.001 (gas[a]) | 14 (I) | 35 (I) | 50 (I) | 51 (V) | 18 (V + VI) | 125 (V + VI) | 187 (VI + V[?]) |
|---|---|---|---|---|---|---|---|---|
| | | | | Assignment (cm$^{-1}$) | | | | |
| $\nu_2(E)$ | 1526 | ~1540(vw, bd)[c] | 1542(vw, bd) | 1551(vw, bd) | 1536(vw, bd) | 1543(vw, bd) | 1544(vw, bd) | 1554(vw, bd) |
| | | ($\omega = 72$)[d] | ($\omega = 72$) | ($\omega = 72$) | ($\omega = 72$) | ($\omega = 70$) | ($\omega = 72$) | ($\omega = 62$) |
| $\nu_4(F_2)$ | 1306.2 | —[e] | —[e] | —[e] | —[e] | —[e] | —[e] | —[e] |
| | — | — | — | — | — | ~2991(vw(sh)) | ~2996(w(sh)) | ~3020(w(sh)) |
| | | | | | | | | ~3035(w(sh)) |
| $\nu_1(A_1)$ | 2914.2(s) | 2926(s) | 2949(s) | 2963(s) | 2966(s) | 3031(s) | 3020(s) | 3070(s) |
| | | ($\omega = 10$) | ($\omega = 10.5$) | ($\omega = 12$) | ($\omega = 14$) | ($\omega = 25$) | ($\omega = 26$) | ($\omega = 36$) |
| | — | — | — | — | — | — | — | ~3139(vw(sh)) |
| | | | | | | | | ~3172(vw(sh)) |
| $\nu_3(F_2)$ | 3020.3(w) | 3058(w, bd) | 3066(w, bd) | 3074(w, bd) | 3086(w, bd) | 3141(w, bd) | 3147(w, bd) | 3199(w, bd) |
| | | ($\omega = 36$) | ($\omega = 64$) | ($\omega = 64$) | ($\omega = 32$) | ($\omega = 32$) | ($\omega = 40$) | ($\omega = 31$) |
| | — | — | — | ~3222(vw, bd) | ~3222(vw, bd) | — | — | — |

[a] From Herzberg (1945).
[b] Measurement accuracy is ±4 cm$^{-1}$ for weak and broad bands, ±1 cm$^{-1}$ for strong and sharp bands.
[c] Abbreviations: s, strong; w, weak; vw, very weak; m, medium; b, broad; sh, shoulder.
[d] $\omega = \frac{1}{2}$ band.
[e] Raman bands could not be measured because of low intensities and interference from very strong first-order ($F_{2g}$) Raman band of diamond at 1333 ± 1 cm$^{-1}$.

# 1. Organic Molecules

tive of a new phase VI forming for which differentiation between a first-order or a second-order transition was not made.

High-pressure infrared studies have also been done. Xu et al. (1981–1982 have examined methane to pressures of greater than 300 kbar. The $\nu_3$ vibration at 3100 cm$^{-1}$ was followed with pressure, and blue shifts were observed. The combination bands at 4200 and 4300 cm$^{-1}$ also demonstrated a blue shift. The fundamental $\nu_3$ vibration showed a decrease in intensity with pressure, while combination bands showed an increase, with a new combination band appearing at 4585 cm$^{-1}$ at the phase transition pressure 50 kbar. Figures 18 and 19 show the pressure dependencies of $\nu_3$ and the combination bands, respectively.

## h. Miscellaneous Organic Compounds

*(1) Adamantanone and Adamantane* Hara et al. (1980) performed x-ray studies of adamantane in the DAC and confirmed the existence of a pressure-induced phase transition from a face-cen-

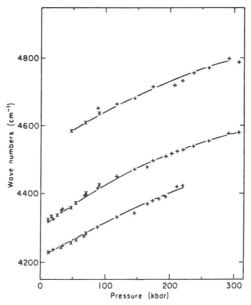

**Fig. 18.** Pressure shift of $\nu_3$ vibration in solid methane. [From Xu et al. (1981–1982).]

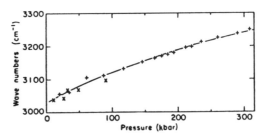

*Fig. 19.* Pressure shift of combination bands in solid methane. [From Xu *et al.* (1981–1982).]

tered cubic structure to a tetragonal structure at 8 kbar and 28°C. The transition was similar to that occurring in adamantanone (see Ito, 1973). Table VI summarizes the crystallographic data.

*(2) Ammonium Salts of Organic Acids* Hamann *et al.* (1973) have examined ammonium acetate and ammonium formate in the DAC with pressures of up to ~50 kbar and used infrared techniques to determine changes.

Ammonium acetate undergoes a phase change at 4.5 kbar at room temperature that is identified by the changes in the IR spectra with pressure. A second phase occurs at ~15 kbar, and changes in the infrared spectra served as the diagnostic technique. Figure 20 shows these data.

Ammonium formate shows a phase change at 11.8 kbar at 25°C and has a triple point at 75°C and 12.75 kbar. Again, infrared spectroscopy was used to identify phase changes.

*Table VI*

*Crystallographic Data for Various Phases of Adamantonone and Adamantane*

| Compound | Pressure | Structure | Cell parameters | | |
|---|---|---|---|---|---|
| | | | $a$ (Å) | $c$ (Å) | $\rho$ (g cm$^{-3}$) |
| Adamantanone | Low pressure | Face-centered cubic | 9.56 | — | 1.14 |
| | High pressure | Tetragonal | 7.15 | 7.82 | 1.25 |
| Adamantane | Low pressure | Face-centered cubic | 9.42 | — | 1.08 |
| | High pressure | Tetragonal | 6.54 | 8.86 | 1.20 |

# 1. Organic Molecules

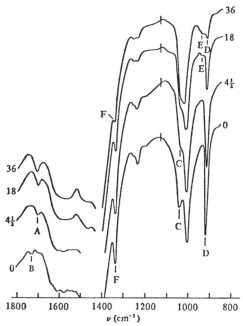

**Fig. 20.** Infrared spectra of ammonium acetate in the high-pressure diamond cell. The numbers on the curves indicate the average pressures $\bar{p}$ in kilobar. The 0-kbar spectrum shows phase I, with peaks B at 1730 cm$^{-1}$ and D at 920 cm$^{-1}$. The $4\frac{1}{2}$-kbar spectrum shows phase II, with a new peak A at 1710 cm$^{-1}$ but peak B at 1730 cm$^{-1}$ removed. The spectra at 18 and 36 kbar show phase III, with a new peak E at 950 cm$^{-1}$ and a relative weakening of the peak D at 920 cm$^{-1}$. [From Hamann et al. (1973).]

*(3) 1,3,5-Trioxane* Hamann et al. (1973) found that a phase change occurred for 1,3,5-trioxane at 22.4 kbar on compression and 12.5 kbar on decompression, indicating that the transition is sluggish. The work confirmed earlier work of Brasch et al. (1970).

*(4) 1,3,5-Trithiane* Hamann et al. (1973) found no phase transition up to a pressure of 50 kbar.

*(5) Alkyl-Substituted Ammonium Halides* Hamann (1975) studied a number of alkyl-substituted ammonium halides in a DAC under pressure and observed infrared evidence for numerous phase changes. Table VII summarizes these data.

### Table VII

*Phase Changes in Several Alkyl-Substituted Ammonium Halides[a]*

| Compound | Type of change | Compound | Type of change |
|---|---|---|---|
| $CH_3NH_3Cl$ | b | $(n-C_3H_7)NH_3Cl$ | b |
| $(CH_3)_2NH_2Cl$ | a | $(n-C_3H_7)_2NH_2Cl$ | b |
| $(CH_3)_3NHCl$ | a | $(n-C_3H_7)_3NHCl$ | b |
| $(CH_3)_4NCl$ | a | $(n-C_3H_7)_4NCl$ | b |
| $(CH_3)_4NBr$ | a | $(n-C_3H_7)_4NBr$ | b |
| $(CH_3)_4NI$ | a | $(n-C_3H_7)_4NI$ | b |
| $(C_2H_5)NH_3Cl$ | a | Cyclohexyl $NH_3Cl$ | a |
| $(C_2H_5)_2NH_2Cl$ | b | $(Cyclohexyl)_2NH_2Cl$ | a |
| $(C_2H_5)_3NHCl$ | a | $CH_3NH_3NO_3$ | a |
| $(C_2H_5)_4NCl$ | b | Propargyl ammonium chloride | a |
| $(C_2H_5)_4NBr$ | a | Hydroxyl ammonium chloride[b] | b |
| $(C_2H_5)_4NI$ | a | | |

[a] Legend: a = new phase, b = strong indication of a new phase.
[b] See Hamann and Linton, 1975a,b.

*(6) Propellants, Explosives, and So On* The Raman and infrared pressure dependency of the propellant HMX, which is octahydro-1,3,5,7-tetranitro-1,3,5,7-tetrazocine, were reported by Goetz *et al.* (1978). The Raman work was done in a DAC at up to 10 kbar, and infrared studies were conducted up to 54 kbar. Four polymorphs of HMX were examined, $\alpha$, $\beta$, $\gamma$, and $\delta$. Beta-HMX was stable to 54 kbar, and $\alpha$-HMX to 42 kbar. Gamma-HMX converts to $\beta$-HMX at 5.5 kbar, and $\delta$-HMX converts to a mixture of $\alpha$- and $\beta$-HMX at a pressure of 0.5 kbar. An inverse relationship was found to exist between temperature and pressure effects; the polymorph most stable at low temperature and having the smallest volume ($\beta$-HMX) is the most stable at high pressure. Figure 21 shows the structural formula of HMX. Figure 22 summarizes the temperature- and pressure-induced phase transitions of HMX. Figure 23 shows pressure-induced spectral changes in the Raman spectra of $\gamma$-HMX, and Fig. 24 shows the infrared spectra of the various phases from 600 to 1800 cm$^{-1}$. Karpowitz and Brill (1983) have determined the kinetic data for some of these phase transitions.

*(7) Pyrazine* Pyrazine was discussed in Chapter 5.

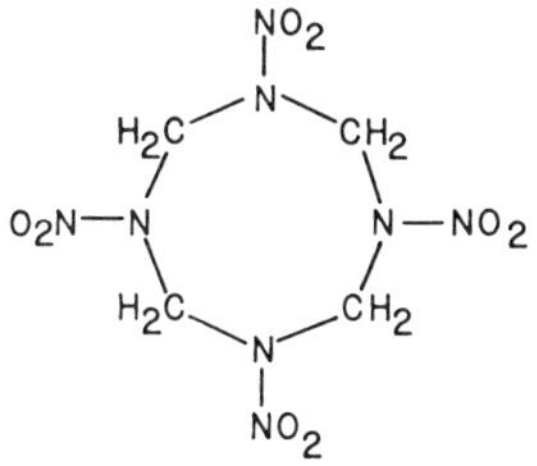

# 1. Organic Molecules

**Fig. 21.** Structural formula of HMX. [Reprinted with permission from Goetz *et al.* (1978). Copyright 1978 American Chemical Society.]

*(8) Dihalocyclohexanes* Klaeboe and Woldback (1978) studied the FIR spectra of six 1,4-dihalocyclohexanes. Pressures of up to 50 kbar were attained in a DAC. Results were interpreted in terms of various conformers.

*(9) Camphor* There are 11 modifications in the phase diagram of camphor reported in a pressure range of up to 35 kbar and between 0 and 200°C (Bridgman, 1952). Several studies that use infrared and Raman spectroscopy have been made for camphor as a function of $T$ and $p$ (Wilkinson *et al.*, 1981; Medina *et al.*, 1982; and Ramnarine *et al.*, 1981). One carbonyl vibration appears in the ambient phase, which splits into two sharp peaks with increased pressure. The results are explained on the basis of the occurrence of a phase transition. These vibrations decrease in frequency with pressure until a second phase transition occurs whereby the trend is reversed and a blue shift is observed. Combination of infrared and Raman data suggested that at least four molecules per unit cell appear for the lower-pressure phase.

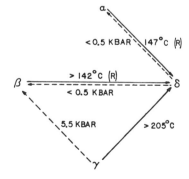

**Fig. 22.** A summary of the temperature- and pressure-induced phase transitions of the $\alpha$, $\beta$, $\gamma$, and $\delta$ polymorphs of HMX. The solid lines are transitions with slow heating, and the dashed lines are transitions with pressure. (R) indicates reversibility with cooling. [Reprinted with permission from Goetz *et al.* (1978). Copyright 1978 American Chemical Society.]

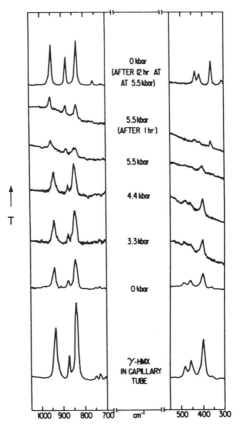

**Fig. 23.** Pressure-induced spectral changes in γ-HMX. A γ- to β-HMX transition occurs at 5.5 kbar. [Reprinted with permission from Goetz *et al.* (1978). Copyright 1978 American Chemical Society.]

*(10) Hydrogen-Bonded Organic Compounds* Benson and Drickamer (1956) examined several organic compounds containing hydrogen under pressure. The systems were studied in $CS_2$ and $CFCl_3$. The vibrations $\nu_{NH}$, $\nu_{SH}$, $\nu_{CH}$, $\nu_{OH}$, and $3\nu_{CF}$(stretch) were followed with pressure. Red shifts were observed for the OH vibration. The CH stretching vibration in $CHCl_3$ and the symmetrical stretching in $CH_2Cl_2$ shows red shifts in $CS_2$, but in $CFCl_3$ they show a blue shift after an initial red shift. The antisymmetric $CH_2$ vibration in $CH_2Cl_2$ shifts blue in both solvents. The $3\nu_{CF}$ stretch shifts to the red. The $\nu_{NH}$ and $\nu_{SH}$ vibrations show initial red shifts followed

1. Organic Molecules

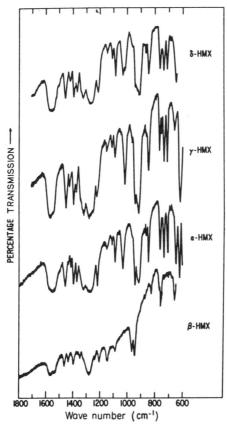

***Fig. 24.*** Infrared spectra (600–1700 cm$^{-1}$) of $\alpha$-, $\beta$-, $\gamma$-, and $\delta$-HMX. Spectra were recorded on neat samples in a diamond anvil cell. [Reprinted with permission from Goetz *et al.* (1978). Copyright 1978 American Chemical Society.]

by a blue shift. Moon and Drickamer (1974) extended this work to a series of organic solids with hydrogen bonds. They studied the effects of pressure on the $\nu_{OH}$ and $\nu_{NH}$ vibrations up to 125 kbar. The NH and OH vibrations shift to lower frequency and broaden substantially. In some cases, the direction reverses after shifts of 200–300 cm$^{-1}$. The $\nu_{CH}$ vibration shows a blue shift with a slight broadening with increasing pressure. Tables VIII and IX summarize the results. The results were explained on the basis of Lippincott's model (Lippincott and Schroeder, 1955) with the addition of a repulsion term.

## Table VIII

H-Bonded O–H(N–H) Stretching Vibration $(Cm^{-1})$[a]

| Compound  P (kbar): | 0 | 25 | 50 | 75 | 100 | 125 | 150 |
|---|---|---|---|---|---|---|---|
| Phenol | 3393 | −54 | −84 | −105 | −120 | (−129) | |
| o-Chlorophenol | 3311 | −95 | −155 | −178 | −174 | (−143) | |
| o-Fluorophenol | 3280 | −114 | −159 | −164 | −151 | (−124) | |
| 2,3-Dichlorophenol | 3398 | −88 | −139 | −179 | (−210) | | |
| 2,4-Dichlorophenol | 3430 | −93 | −142 | −165 | −183 | (−199) | |
| 2,6-Dichlorophenol | 3435 | −63 | −105 | −132 | −155 | (−177) | |
| 2,3,4-Trichlorophenol | 3502 | −26 | −38 | −46 | −53 | −60 | (−65) |
| 2,3,5-Trichlorophenol | 3502 | −26 | −45 | −61 | −76 | −93 | (−106) |
|  | 3442 | −61 | −101 | −124 | −145 | −166 | (−182) |
|  | 3329 | −124 | (−209) | | | | |
| 2,3,6-Trichlorophenol | 3508 | −25 | (−50) | | | | |
|  | 3470 | −52 | (−91) | | | | |
|  | 3393 | −90 | (−156) | | | | |
| Catechol | 3450 | −59 | −98 | −123 | −146 | −166 | (−183) |
|  | 3427 | −154 | −189 | −214 | −237 | −256 | (−272) |
| Resorcinol | 3223 | −153 | −196 | −139 | (−74) | | |
| Hydroquinone | 3234 | −123 | −214 | −268 | −307 | (−344) | |
| N-Methylacetamide | 3283 | −22 | −32 | −37 | −35 | −29 | (−19) |
| Aniline | 3430 | −7 | −14 | −19 | −23 | −26 | (−30) |
|  | 3350 | −10 | −12 | −9 | −6 | −3 | (+1) |
|  | 3231 | −10 | −20 | −31 | −42 | −53 | (−63) |
| Boric acid | 3228 | −117 | −173 | −186 | −170 | (−130) | |

[a] Red shifts indicated as negative numbers.

## Table IX

C–H Stretching Vibration $(Cm^{-1})$[a]

| Compound  p (kbar): | 0 | 25 | 50 | 75 | 100 | 125 | 150 |
|---|---|---|---|---|---|---|---|
| Benzene | 3092 | 14 | 29 | 46 | 63 | 85 | (106) |
|  | 3036 | 14 | 28 | 45 | 60 | (76) | |
|  | (3024) | 11 | 21 | 31 | 42 | 54 | (66) |
| 2,3-Dichlorophenol | 3093 | 13 | 28 | 46 | 66 | (85) | |
| 2,4-Dichlorophenol | 3093 | 17 | 37 | 56 | 74 | 92 | (110) |
| 2,6-Dichlorophenol | 3072 | 19 | 33 | 45 | 57 | 70 | (84) |
| 2,3,4-Trichlorophenol | (3107) | 15 | 31 | 46 | 62 | 79 | (96) |
|  | 3077 | 14 | 27 | 39 | 51 | 63 | (75) |
| 2,3,5-Trichlorophenol | 3088 | 23 | 41 | 55 | 71 | 87 | (104) |

[a] Blue shifts indicated as positive numbers.

**Table X**

*Phase Changes in Several Organic Solids Resulting with Pressure*[a]

| Compound | Type of change |
|---|---|
| Acetamide | a |
| p-Benzoquinone | a, b |
| Guanidine sulfate | a |
| o-Nitrophenol | b |
| Phthalic anhydride | c |
| Salicylic acid | c |
| Thiourea | c |
| Urea | c |
| Vanillin | b |

[a] Legend: a = confirmation of a previously reported phase, b = new phase, and c = new phase strongly indicated.

Hamann and Linton (1975a) studied several organic acids in a DAC with pressure of up to 40 kbar. The hydrogen-bonded stretch and bend were followed in the infrared. The acids studied were salicyclic, adipic, succinic, mellitic, cinnamic, and maleic. Other compounds examined were 2,2′,4,4′-tetramethylpentane-3-one oxime and potassium hydrogen(deuterium)di-o-nitrobenzoate. Stretching modes showed a decrease with pressure. Bending vibrations all increased in frequency with pressure. The only exceptions were compounds with strong symmetrical hydrogen bonds, in which the stretches increased in frequency with pressure (in this study the potassium hydrogen(deuterium)di-o-nitrobenzoate).

*(11) Other Compounds* Hamann and Linton (1975b) examined other miscellaneous organic compounds under pressure of up to 50 kbar in a DAC and found or confirmed several new pressure phases. Infrared spectroscopy was used to identify the changes. Table X tabulates the results.

## 2. BIOLOGICAL COMPOUNDS

What was true for solid organic molecules is also true for solid biological molecules, perhaps to a greater extent. Vibrational spectroscopy at elevated pressures of solid biochemical systems has been almost totally neglected. There are excellent reasons to examine biological compounds under pressure. Perhaps the greatest reason is concerned with the fact that a large portion of the biomass on earth exists under the pressure of the oceans. How do organisms adapt to the pressure environment? Since a good deal of these systems involve $H_2O$ solutions, studies of biochemicals in aqueous solutions under pressure that use a DAC could prove worthwhile. This area remains a virgin field.

For review articles on pressure effects on biochemical systems in general, we refer the reader to the articles listed at the end of this chapter.

### a. Bacteriorhodopsin

Crespi and Ferraro (1979) studied freeze-dried bacteriorhodopsin (bR) in a DAC under pressure. Visible spectroscopy was used as the

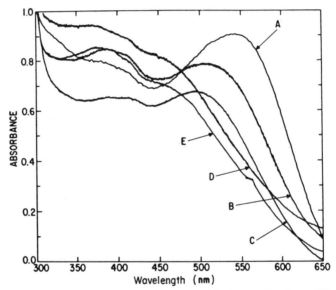

***Fig. 25.*** Optical spectra of freeze-dried natural hydrogen abundance ($^1$H) bacteriorhodopsin (bR), showing the sequential color changes involved in going from the starting material at no pressure (A), blue-shifted bR at pressures of 8 kbar (B) and 20 kbar (C), and bleached bR on release of pressure (D). After standing 3 hr under ambient conditions (E), the bleached bR showed a slight recovery of absorption at 560 nm. Bleached bR is also observed when the entire sequence of pressure application and release is conducted in the dark. [From Crespi and Ferraro (1979).]

## 2. Biological Compounds

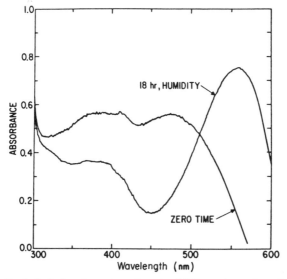

*Fig. 26.* Bacteriorhodopsin recovers its natural 560-nm absorption when the pressure-bleached material is exposed to 100% relative humidity at room temperature at ambient pressure. This experiment was done with fully deuterated ($^2$H) bR. The same result is obtained with $^1$H material. [From Crespi and Ferraro (1979).]

diagnostic tool. Figure 25 shows the response of bR to pressure. As pressure is applied, the absorption maximum shifts to the blue from 540 to 500 nm, and an absorption grows at 390–400 nm. At 3 kbar with release of pressure to this point, there appears to be instant and extensive bleaching of the sample. Similar results were obtained when the bR was exchanged with $D_2O$. It was found that bR recovers its natural 560-nm absorption when the pressure-bleached material is exposed to 100% relative humidity at room temperature (see Fig. 26). The results gave insight into the active site structure in bR and its mechanism of action.

### b. α-glycine

Medina *et al.* (1980) have conducted Raman studies of α-glycine as a function of $T$ and $p$. Pressure dependencies of the most resolved bands in α-glycine were determined, and all of these demonstrate a blue shift.

## REFERENCES

Adams, D. M., and Appleby, R. (1977). *J. Chem. Soc. Faraday Trans. 2*, 1896.
Adams, D. M., and Ejejiuba, I. O. C. (1981). *J. Chem. Soc. Faraday Trans.* **77**, 851.
Adams, D. M., and Sharma, S. K. (1976). *J. Chem. Soc. Dalton Trans.* 2424.
Adams, D. M., Shaw, A. C., Mackenzie, G. A., and Pawley, G. S. (1980). *J. Phys. Chem. Solids* **47**, 149.
Anderson, A., and Walmsley, S. H. (1965). *Mol. Phys.* **9**, 1.
Benson, A. M., and Drickamer, H. G. (1956). *Faraday Discuss.* 39.
Bockelmann, H. K., and Sherman, W. F., eds. (1980). *Adv. Infrared Raman Spectros.* **6**, 280.
Brasch, J. W., Melveger, A. J., Lippincott, E. R., and Hamann, S. D. (1970). *Appl. Spectrosc.* **24**, 184.
Bridgman, P. W. (1952). "The Physics of High Pressures." Bell and Son, London.
Constantino, M. S., and Daniels, W. B. (1975). *J. Chem. Phys.* **62**, 764.
Crespi, H. L., and Ferraro, J. R. (1979). *Biochem. Biophys. Res. Commun.* **91**, 575.
Dows, D. A., Hse, L., Mitra, S. S., Brafman, O., Hayek, M., Daniels, W. B., and Crawford, R. K. (1973). *Chem. Phys. Lett.* **22**, 595.
Ellenson, W. D., and Nicol, M. (1974). *J. Chem. Phys.* **61**, 1380.
Goetz, F., Brill, T. B., and Ferraro, J. R. (1978). *J. Phys. Chem.* **82**, 1912.
Hamann, S. D. (1975). *High Temp.–High Pressures* **7**, 177.
Hamann, S. D. (1976). *High Temp.–High Pressures* **8**, 317.
Hamann, S. D. (1978). *High Temp.–High Pressures* **10**, 97.
Hamann, S. D., and Linton, M. (1975a). *Aust. J. Chem.* **28**, 2567.
Hamann, S. D., and Linton, M. (1975b). *High Temp.–High Pressures* **7**, 165.
Hamann, S. D., Linton, M., and Pistorious, C. W. F. T. (1973). *High Temp.–High Pressures* **5**, 575.
Hara, K., Osugi, J., Taniguchi, Y., and Suzuki, K. (1980). *High Temp.–High Pressures* **12**, 221.
Hazen, R. M., Mao, H. K., Finger, L. W., and Bell, P. M. (1979–1980). *Annu. Rep. Geophys. Lab., Washington, D.C.* **79**, 348.
Herzberg, G. (1945). "Molecular Spectra and Molecular Structure," Vol. 2. Van Nostrand, New York.
Ito, T. (1973). *Acta Crystallogr. B.* **29**, 364.
Karpowitz, R. J., and Brill, T. B. (1983). *Appl. Spectrosc.* **37**, 79.
Klaeboe, P., and Woldback, T. (1978). *Appl. Spectrosc.* **32**, 588.
Lippincott, E. R., and Schroeder, R. (1955). *J. Chem. Phys.* **23**, 1099.
Mammone, J. F., Sharma, S. K., and Nicol, M. (1980). *J. Phys. Chem.* **84**, 3130.
Medina, J. A., Sherman, W. F., and Wilkinson, G. R. (1980). *Proc. Int. Conf. Raman Spectrosc., 7th, Ottawa, Canada,* p. 112. North-Holland Publ., Amsterdam.
Medina, J. A., Sherman, W. F., Stadtmuller, A. A., and Wilkinson, G. R. (1982). *Spectrochim. Acta, Part A* **38**, 483.
Moon, S. H., and Drickamer, H. G. (1974). *J. Chem. Phys.* **61**, 48.
Ramnarine, R., Sherman, W. F., and Wilkinson, G. R. (1981). *Infrared Phys.* **21**, 391.
Sharma, S. K., Mao, H. K., and Bell, P. M. (1979–1980). *Annu. Rep. Geophys. Lab., Washington, D.C.* **79**, 351.
Sherman, W. F., and Smulovitch, P. P. (1970). *J. Chem. Phys.* **52**, 5187.
Wilkinson, G. R., Medina, J. A. and Sherman, W. F. (1981). *J. Raman Spectrosc.* **10**, 155.

Wu, C. F., and Jura, G. (1973). *Spectrochim. Acta, Part A* **30,** 797.
Xu, J., Hoering, T. C., Mao, H. K., Wong, P., and Bell, P. M. (1981–1982). *Annu. Rep. Geophys. Lab., Washington, D.C.* **81,** 389.

## REVIEW ARTICLES

### Pressure Effects on Organic Compounds

Bertie, J. E. (1980). Infrared spectra of solids at normal and high pressures. *In* "Analysis and Application of FT-IR to Molecular and Biological Systems" (J. R. Durig, ed.), Vol. 57, pp. 467–500. D. Reidel, Dordrecht, Netherlands.
leNoble, W. J. (1978). Organic model reactions under pressure. *In* "High-Pressure Chemistry" (H. Kelm, ed.), pp. 325–345. D. Reidel, Dordrecht, Netherlands.
leNoble, W. J. (1978). Organic problem reactions pressure. *In* "High-Pressure Chemistry" (1978). (H. Kelm, ed.), pp. 345–363. D. Reidel, Dordrecht, Netherlands.

### High-Pressure Effects on Biological Compounds

Heremans, K. (1980). Pressure effects on biochemical systems. *In* "High Pressure Chemistry" (H. Kelm, ed.), p. 467. D. Reidel, Dordrecht, Netherlands.
Hochachka, P. W. (1976). "Biochemistry at Depth." Pergamon, Oxford.
Johnson, F. H., Eyring, H., and Stover, B. J. (1974). "The Theory of Rate Processes in Biological Medicine." Wiley (Interscience), New York.
MacDonald, A. G. (1975). "Physiological Aspects of Deep Sea Biology." Cambridge Univ. Press, Cambridge.
Marquis, R. E. (1976). *Adv. Mar. Microbial Physiol.* **14,** 159.
Sleigh, M. A., and MacDonald, A. G., eds. (1972). "The Effects of Pressure on Organisms." Cambridge Univ. Press, Cambridge.
Somero, G. N., and Hochachka, P. W. (1976). *In* "Adaptations to Environment" (R. C. Newell, ed.). Butterworth, London.
Zimmerman, A. M., ed. (1970). "High Pressure Effects on Cellular Processes." Academic Press, New York.

## BIBLIOGRAPHY

Fiedler, K. M., Wunder, S. L., Priest, R. G., and Schnur, J. M. (1982). *J. Chem. Phys.* **76,** 5541.
Takemura, T. (1979). *Polym. Prepr. Am. Chem. Soc., Div. Polym. Chem.* **20,** 270.
Tanaka, H., and Takemura, T. (1980). *Polym. J.* **12,** 350.
Wunder, S. L., Cavatorta, F., Priest, R. G., Schoer, P. E., Sheridan, J. P., and Schnur, J. M. (1979). *Polym. Prepr. Am. Chem. Soc., Div. Polym. Chem.* **20,** 776.
Wunder, S. L. (1981). *Macromol. Rev.* **14,** 1024.

CHAPTER 7

# SPECIAL APPLICATIONS

The versatility of the DAC is demonstrated by the number of applications that have surfaced with time. The capability of reaching megabar pressures with the ultra-high-pressure DAC has generated considerable interest in the field of geochemistry and geophysics. It has also been found useful in electrical conductor research, a new, developing field. The DAC has been found useful in the study of lubricants. Since the DAC can serve as a microanalytical technique, it has generated interest in the area of forensic science. These applications will be discussed in this chapter in an introductory fashion. The reader is referred to the original articles for a comprehensive treatment of each subject.

## 1. GEOCHEMICAL AND GEOPHYSICAL APPLICATIONS

One of the most important applications of the DAC has been in the area of geophysics and geochemistry (Bassett, 1979; Mao and Bell, 1978). The development of the ultra-high-pressure DAC has been responsible for this interest (Piermarini and Block, 1975; Mao et al., 1978; Barnett et al., 1973; Bell, 1983). Results obtained from the visible region of the electromagnetic spectrum (Mao 1973a,b; Mao and Bell, 1972a–c) and to a lesser extent from the infrared region through Raman and Brilluoin scattering (Whitfield *et al.*, 1976; Bassett and Brody, 1977; Sharma *et al.*, 1980), combined with data from electrical conductivity (Mao and Bell, 1977, 1981, 1972a–c; Mao 1973a–c; Block and Piermarini, 1976; Walling and Ferraro, 1978), Mössbauer spectroscopy (Mao *et al.*, 1977; Huggins *et al.*, 1975), and x-ray spectroscopy (Merrill and Bassett, 1974; Bassett and Takahashi, 1974; Bassett *et al.*, 1967, 1974, 1979; Piermarini and Weir, 1962; Weir *et al.*, 1969; Hazen, 1977a,b), have allowed the geophysicist and geochemist to accumulate laboratory data on minerals under conditions as they exist in the interior of the earth. Experimental conditions simulating the conditions of temperature and pressure existing at the core–mantle discontinuity have become possible and have served to provide more realistic conclusions concerning the nature of the earth's interior and the effects of pressure and temperature on the important optical, electrical, magnetic, thermal, and other properties of the earth.

Virtually all of the direct information about the interior of the earth has come from observations of the propagation of very-low-frequency elastic sound waves generated by earthquakes. One class of waves are the surface waves, which are guided by density and velocity layering at or near the surface and are important in the elucidation of crystal and upper-mantle structures. Of greater interest are the body waves, which penetrate the interior. The latter waves are of two types called the P waves (compressional) and S waves (shear). The designations P and S stand for the order of arrival at a measuring station. The P (primary) waves arrive first, while the S (secondary) waves arrive later. Complementary information has also come from wave-propagation measurements on rocks (Birch, 1960, 1961), magnetic variation data, and shock-wave data of silicates (Hughes and McQueen, 1958; Altschuler and Kormer, 1961; Lombard, 1961; Wackerle, 1962). Some of these data are illustrated in Figs. 1 and 2, and Fig. 1 (Ahrens, 1980) shows the seismic velocity distribution in the mantle. Figure 2a illustrates the electrical conductivity as a function of the earth's depth (from magnetic variation data), and Fig. 2b shows the variation of the temperature with depth (obtained from conductivity data).

To this background body of information we can now add new laboratory data obtained with the DAC. Before we discuss these data, let us examine previous results of pressure obtained on materials related to minerals such as inorganic and coordination com-

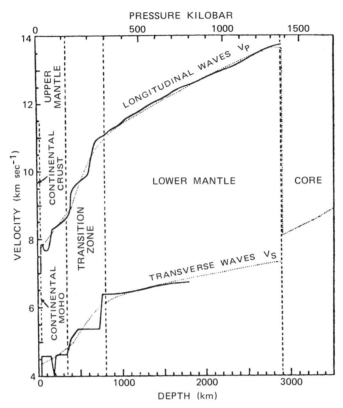

*Fig. 1.* Seismic velocity distributions in the mantle. P waves, solid curves; S waves, dotted curves. [From Mao (1973c).]

pounds. In Chapters 4 and 5 we discussed the effects of high pressure on the vibrational transitions of inorganic and coordination compounds. Drickamer and Frank (1973) have discussed similar effects on electronic transitions. From this foundation we can reasonably predict what we would expect from pressure experiments of minerals. We would expect that the following factors would be affected:

(a) Charge-transfer transitions would be affected by pressure, and this would be a dominant process occurring in the earth's interior. A red shift accompanied by an increase in intensity would generally be observed.

# 1. Geochemical and Geophysical Applications

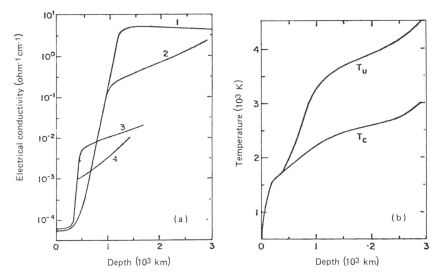

*Fig. 2.* (a) Electrical conductivity of mantle. (b) Temperature within mantle estimated from electrical conductivity. [From Mao (1973c).]

(b) The electrical resistance would show a decrease, and consequently electrical conductivity would increase with pressure (see Fig. 2a).

(c) Magnetic properties would be affected because spin states are pressure labile. For example, spin states could change. High-spin states could be converted to low- or intermediate-spin states.

(d) Oxidation states could change. In general, reduction or disproportionation of oxidation states would be possible.

(e) Coordination numbers around a central atom would increase.

(f) More compact, efficient, packed structures would be stable. Phases with higher densities and lower volume would be found.

(g) Phase transitions would occur, providing phases that correspond to the criteria of (e) and (f).

These factors would affect various physical properties of minerals existing in the interior. Factor (a) would be instrumental in changing the optical and thermal properties. Factor (b) would modify electrical properties. Factors (c) and (d) would induce changes in the magnetic properties. From these considerations it is obvious that surface minerals would not survive in the earth's interior. Any relevant model of the earth must deal with the state of the minerals stable at conditions of temperature and pressure occurring in the interior.

## a. Structural Response of Minerals to Changes in Temperature and Pressure

Thus far, we have only considered pressure effects and neglected temperature effects. Since both parameters are elevated together in the earth, it is important that studies of various phases be directed toward simultaneous changes in both $T$ and $p$. Some research in this direction has been started by Hazen (1977a,b). It has been found that the effect of temperature on the transition pressures of Si(IV) to Si(VI) structures is small. In the conversion of coesite to stishovite, the observed pressure dependency on temperature is 0.01 kbar $°C^{-1}$ (Yagi and Akimoto, 1976). Hazen (1977a,b) and Hazen and Prewitt (1977) discuss changes in the ratios of polyhedral sizes in oxygen-based minerals caused by changes in $T$ and $p$ and composition. Two empirical relationships between bond expansion compression and other bonding variables were defined:

$$\bar{\alpha} = 32.9(0.75 - Z/q) \times 10^{-6} \ °C^{-1}, \tag{1}$$

$$\bar{\beta} = 37.0(d_0^3/Z) \times 10^{-6} \ \text{kbar}, \tag{2}$$

where $Z$ represents cation charge, $q$ coordination number, $d_0$ the mean polyhedral metal–oxygen distance at 1 atm and 23°C, $\bar{\alpha}$ the mean coefficient of linear expansion from 23 to 1000°C, and $\bar{\beta}$ the mean coefficient of linear compression to 100 kbar. For many cation polyhedra such as $Si^{4+}$(IV), $Al^{3+}$(VI), $Be^{2+}$(IV), $Mg^{2+}$(VI), $Fe^{2+}$(VI), $Ni^{2+}$(VI), and $Co^{2+}$(VI), the observed $\bar{\alpha}/\bar{\beta}$ ratios are approximately constant. For minerals containing these cations, the bond-length changes during thermal expansion will be the inverse of those during compression. From these relationships it has become possible to predict bond compressions and expansions for other cation polyhedra and estimate structural variations with temperature and pressure for many oxygen-based minerals, a necessary requirement for better comprehension of the effects of temperature and pressure.

Although visual and electrical observations have been made simultaneously at high pressure and varying temperature in the DAC,

# 1. Geochemical and Geophysical Applications

more research in this area must be accomplished. Other experimental measurements are needed with simultaneous changes in temperature and pressure. Certainly, this area of research could prove very fruitful in elucidating the secrets of the interior of the earth.

## b. General Effects of Pressure on Different Types of Mineral Lattices

Prior to examining experimental results of pressure effects on minerals, it is advisable to investigate the general overall effects of pressure on the primary types of mineral lattices existing. Three types of solid morphology can be considered for mantle minerals (Hazen and Finger, 1978; Hazen and Prewitt, 1977). These are layer, coordination, and metallic lattices.

*(1) Layer Lattices* Examples of these types of lattices are graphite and boron nitride. In general, in these types of lattices intermolecular distance (between layers) will decrease greatly, whereas intramolecular distances will increase slightly with a pressure increase. Densities will increase, and coordination numbers (CNs) will tend to increase. Table I will illustrate.

*(2) Coordination Lattices* Typical examples of coordination lattices are coesite and quartz. Here it is found that with an increase in pressure, bond distances (between heteroatoms) increase, whereas those between homoatoms decrease (lattice modes). Coordination numbers and density increase. Table II will illustrate these points.

*(3) Metallic Lattices* In metallic lattices the coordination number is 12 and is at its maximum. Pressure will contract lattices, causing a delocalization of electrons, and intermolecular distances will decrease.

### Table I
*Effects of Pressure on Layer Lattice Compounds*

| Example | CN | Interdistances | Intradistances | $D$ |
|---|---|---|---|---|
| Graphite $\rightarrow$ diamond | $3 \rightarrow 4$ | Large decrease | Slight increase | Increase |
| BN $\rightarrow$ cubic (zinc blende) (B-12 type) | 4 | Large decrease | Slight increase | Increase |

**Table II**

*Effects of Pressure on Coordination Lattice Compounds*

| Example | CN | $d_{ZnS(ZnSe)}$ | $d_{Zn-Zn}$ | D |
|---|---|---|---|---|
| ZnS, ZnSe (zinc blende) → NaCl | 4 → 6 | Increase | Decrease | Increase |

| Example | CN | $d_{SiO}$ | $d_{Si-Si}$ | |
|---|---|---|---|---|
| Coesite → shishovite | 4 → 6 | Increase | Decrease | Increase |
| quartz → coesite | 4 → 4 | Increase | Decrease | Slight increase |
| RbCl → CsCl | 6 → 8 | Increase | Decrease | Increase |

### c. Newer Laboratory Data on Minerals

Birch (1952) postulated that the high-pressure phases of ferromagnesium silicates that exist in the lower mantle must be a mixture of dense oxides. He based these conclusions on the need to account for the rapid increase in seismic velocities with depth observed in the upper mantle. Later shock-wave experiments verified these conclusions (McQueen and Marsh, 1966); see Table III (Ahrens, 1980).

*(1) Transformation of Silicate Minerals with Pressure* Ringwood and co-workers (Ringwood, 1962, 1963, 1966, 1975; Ringwood and Major, 1966) studied various phase transformations in mantle silicates under pressure and divided the transformation into two groups: (1) changes occurring in packing of polyhedra with no change in cation coordination (e.g., olivine → γ → spinel, in which Si remains at coordination number 4); packing in this type is inefficient because $SiO_4$ units in $T_d$ geometry share only corners and edges; and (2) changes involving changes in cation coordination (at least one cation) [e.g., Si(IV) → Si(VI) → spinel → stishovite]. Packing restrictions are reduced in the $O_h$ geometry since face sharing becomes possible with the higher coordination number.

Hazen and Finger (1978) discussed the crystal chemistry of Si–O bonds at high pressure. They analyzed results obtained by x-ray measurements at high pressure and confirmed Ringwood's conclusion that two groups of phase transformation took place. Table II summarizes Si(IV) → Si(VI) transformations. A consequence of these studies has led to conclusions that $\frac{2}{3}$ of the mantle transition

### Table III
#### Shock-Induced Phase Changes of Geophysical Importance[a]

| | Low-pressure phase | | High-pressure phase | |
|---|---|---|---|---|
| Material | Density[b] (g cm$^{-3}$) | | Structure structure | Density[b] (g cm$^{-3}$) |
| *Core materials* | | | | |
| Iron, body-centered cubic, Fe | 7.875 | | Hexagonal close packed | 8.32 |
| Iron-nickel (10%), body-centered cubic, Fe$_{0.9}$Ni$_{0.1}$ | 7.884, 7.86 | | Hexagonal close packed | 8.33 |
| Pyrrhotite, Fe$_{0.94}$S | 4.603 | | ? | 5.3–5.5 |
| Pyrite, FeS$_2$ | 4.61 | | ? | 5.3 |
| Wüstite, Fe$_{0.94}$O | 5.50 | | ? | 5.84 |
| 3Fe$_3$Si + FeSi | 7.016 | | Hexagonal close packed | 7.46 |
| Fe$_3$Si + FeSi | 7.646 | | Hexagonal close packed | 8.51 |
| Magnetite, Fe$_3$O$_4$ | 5.12 | | FeO (LS) + Fe$_2$O$_3$ (LS)[c] | 6.4 |
| Hematite, Fe$_2$O$_3$ | 5.00 | | Corundum (LS) | 6.05 |
| *Mantle materials* | | | | |
| Periclase, MgO | 3.58 | | None | |
| Corundum, Al$_2$O$_3$ | 3.98 | | None | |
| Quartz, SiO$_2$ | 2.65 | | Rutile | 3.248 |
| Bronxite, (Mg$_{0.92}$Fe$_{0.08}$)SiO$_3$ | 3.34 | | Hypothetical perovskite | 4.20 |
| Forsterite, Mg$_2$SiO$_4$ | 3.32 | | Hypothetical perovskite + rock salt | 3.93 |
| Dunite, (Mg$_{0.88}$, Fe$_{0.12}$)$_2$SiO$_4$ | 3.55 | | Hypothetical perovskite + rock salt | 4.04 |
| Garnet, (Fe$_{0.79}$, Mg$_{0.14}$, Ca$_{0.04}$, Mn$_{0.03}$)Al$_2$Si$_3$O$_{12}$ | 4.18 | | ? | ~4.4 |
| Spinel, MgAl$_2$O$_4$ | 3.582 | | Orthorhombic | 4.31 |
| Dunite, (Mg$_{0.45}$, Fe$_{0.55}$)$_2$SiO$_4$ | 3.85 | | Hypothetical perovskite + rock salt | 4.62 |
| Calcia, CaO | 3.35 | | CsCl | 3.71 |

[a] Taken in part from Ahrens (1980).
[b] Values are for 1 bar and 25°C.
[c] LS, Fe$^{2+}$ assumed to be in a low-spin orbital configuration.

zone (500–900 km) involves the Si(IV) → Si(VI) transition. Below 900 km all Si is in $O_h$ geometry (CN = 6) or in even greater coordination numbers (Liu, 1978).

Table IV illustrates that spinels transform with pressure to a mixture of oxides (stishovite and metal oxide) (Lui, 1975a,b, 1976). For example, ferrous silicate spinel ($Fe_2SiO_4$) disproportionates at ~250 kbar into wüstite (FeO) and stishovite ($SiO_2$). Clinoferrosilite at similar pressures transforms to FeO and stishovite (Lui, 1976). Ming and Bassett (1975) with a DAC at 250 kbar and 1700°C using a YAG laser converted olivines of varying compositions ($Fo_0 Fa_{100}$, $Fo_{60} Fa_{40}$, $Fo_{80} Fa_{20}$, and $Fo_{100} Fa_0$) and spinel ($Fo_{50} Fa_{50}$ and $Fo_0 Fa_{100}$) to (Mg,Fe)O and $SiO_2$ (stishovite), where $Fo_n = n\%$ fosterite ($Mg_2SO_4$) and $Fa_n = n\%$ fayalite ($Fe_2SO_4$). It has been found that the natural or synthetic olivine passes through an intermediate state ($\gamma$ – spinel) at 50–150 kbar and 1000°C prior to forming the oxides (Ringwood, 1958, 1962, 1963, 1966). Ringwood and Major (1966) converted clinopyroxene (Fe,Mg)$SiO_3$ into (Fe,Mg)$_2SiO_4$ (spinel) and stishovite at 80–180 kbar and 100°C. The reactions can be depicted as

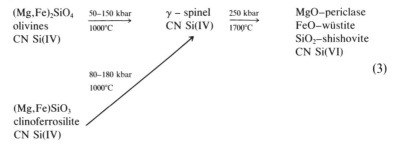

(3)

On the basis of these experiments, it is clear that olivines and pyroxenes convert with pressure and temperature to a mixture of oxides as indicated in the reaction (3).

The experimental work previously discussed led Ming and Bassett (1975) and Bassett (1979) to propose pressure–composition phase diagrams for the $FeO–SiO_2$ and $MgO–SiO_2$ systems as well as for the $MgSiO_3–Al_2O_3$, $MgSiO_3$ and $FeSiO_3$, and $Mg_2SiO_4–Fe_2SiO_4$ systems. Figure 3 illustrates the phase diagram of the latter system. Thus, it has been concluded that perovskite is a prime constituent in the lower mantle of the earth. Below 650 km the mantle mineralogy appears to be consistent with magnesiowüstite ((Mg,Fe)O) and stishovite.

Table IV

High-Pressure $IV_{Si} \rightarrow VI_{Si}$ Transformations

| Formula | $IV_{Si}$ phase | | $VI_{Si}$ phase | Transition $p_T$ (kbar) | Density increase (%)[a] |
|---|---|---|---|---|---|
| $SiO_2$ | Coesite | | Stishovite | $80 \pm 2$ | 50 |
| $Al_2SiO_5$ | Kyanite | | Corundum + stishovite | $160 \pm 20$ | 11 |
| $MgSiO_3$ | Clinoenstatite | $\rightarrow$ [ stishovite + $\beta$-$Mg_2SiO_4$ ] [b] $\rightarrow$ Ilmenite type | | $[170] \rightarrow 250$ | $[10] \rightarrow 10$ |
| $FeSiO_3$ | Clinoferrosilite | $\rightarrow$ [ stishovite + $\gamma$-$Fe_2SiO_4$ ] $\rightarrow$ Wüstite + stishovite | | $[150] \rightarrow 250$ | $[10] \rightarrow 10$ |
| $CaSiO_3$ | Wollastonite | | Perovskite type | 160 | — |
| $ZnSiO_3$ | Zn-pyroxene | | Ilmenite type | 180 | — |
| $Mg_2SiO_4$ | Mg–silicate–spinel | | Perovskite-type + stishovite | 270 | ~10 |
| $Fe_2SiO_4$ | Fe-silicate–spinel | | Wüstite + stishovite | ~250 | ~10 |
| $Ni_2SiO_4$ | Ni-silicate–spinel | | NiO + stishovite | 190 | — |
| $Co_2SiO_4$ | Co-silicate–spinel | | CoO + stishovite | $180 \pm 10$ | — |
| $NaAlSi_2O_6$ | Jadeite | | $NaAlSiO_4$ (calcium ferrite type) + stishovite | 180 | 19 |
| $KAlSi_3O_8$ | Orthoclase | $\rightarrow$ [ kyanite + coesite + $K_2Si_4O_9$ (wadeite type) ] $\rightarrow$ Hollandite type | | $[\sim60] \rightarrow 100$ | $[25] \rightarrow 25$ |
| $Mg_3Al_2Si_3O_{12}$ | Pyrope | | Ilmenite type | $245 \pm 5$ | 7 |

[a] Density increase at 1 atm (23°C).
[b] Brackets indicate transformation from $IV_{Si}$ to mixed $IV_{Si}$ plus $VI_{Si}$ assemblages.

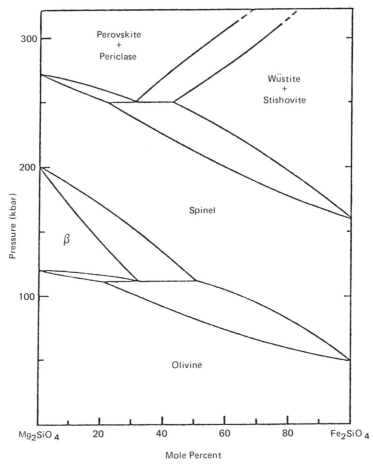

**Fig. 3.** Proposed pressure–composition phase diagram for the olivine compound system $Mg_2SiO_4$–$Fe_2SiO_4$. [From Ruoff (1979).]

*(2) Disproportionation of Iron* Mao and Bell (1971, 1977) and Bell and Mao (1975) have shown that various oxidation states of iron can coexist in equilibrium in the earth. Bell et al. (1976) subjected a synthetic basalt glass (FeO, MgO, $Al_2O_3$, $SiO_2$) to a pressure of 150 kbar and 1000°C and obtained elemental iron and a garnet phase containing ferric iron. Calculations by Mao (1974) showed that the disproportionation of ferrous iron to a mixture of ferric iron and

elemental iron is formed because the volume of the mixture is smaller. Bell and Mao (1975) also found that the spinel phase of fayalite composition disproportionated into stishovite, wüstite ($Fe_{0.91-0.97}O$), and metallic iron. The presence of metallic iron was established by use of an electron microscope and confirmed by Bassett (1979), who used the ion probe method. These results suggest that metallic iron and $Fe^{2+}$ and $Fe^{3+}$ can coexist deep within the mantle and provides a mechanism for forming a metal core from the breakdown of FeO. The possibility exists that some FeO may be dissolved in the molten iron to account for the lower density observed in the core.

*(3) Properties of Oxides with Pressure* The iron oxides are of interest because they modulate the optical, electrical, and thermal properties of the earth. Ferrous oxide is largely transparent to thermal radiation according to Mao and Bell (1975), whereas metallic iron and higher oxides of iron are opaque, which blocks radiative heat transfer (Mao, 1976, 1974).

Mao (1973b) studied the absorption spectra of $Wü_{22}Pe_{88}$ at high pressure and room temperature. Figure 4 depicts the response with pressure. A crystal field absorption at 10,000 $cm^{-1}$ shifts to higher energy, and the absorption edge shifts red. Figure 5 shows the electrical conductivity response to pressure of four magnesiowüstites. It is observed that conductivity increases with pressure and that wüstite fractions show the higher conductivities, as expected.

Continuous changes in the electronic properties of minerals are occurring with pressure and temperature within the earth. If charge transfer bands are shifted into the red, radiative heat transfer will be blocked. Thus, the heat transfer must occur by other pathways, such as electronic or excitonic processes.

It should be noted that phases stable at high pressures assume the highest coordination numbers and are dense. These characteristics allow very efficient packing. Hazen and Finger (1978) reported that high-coordination polyhedra are more compressible than systems in low coordination and that polyhedra tend to become more symmetrical with pressure.

*(4) Conclusions* The results to date can lead to certain conclusions regarding the minerals existing at various depths. The follow-

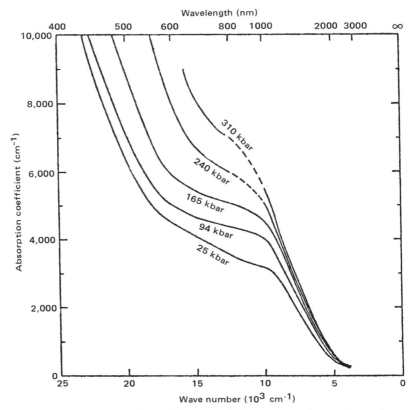

**Fig. 4.** Optical absorption of Wü$_{22}$Pe$_{78}$ at high pressure and room temperature. Dashed curves indicate region in which detector sensitivity is low. [From Mao (1973b).]

ing breakdown is a possible mineral distribution as one proceeds into the earth:

*Upper Mantle* (150–400 km)
    Olivines (Fe,MgSO$_4$)
    Pyroxene (Mg,Fe)SiO$_3$     CN of Si = 4
    Pyrope (Mg$_3$Al$_2$Si$_3$O$_{12}$)

*Transition zone* (400–1000 km)
    Perovskite (MgSiO$_3$)     CN of Si = 4
    Spinel (M$_2$SiO$_4$, M = Mg, Fe, Ni, Co)     CN of Si = 4
    Ileminite (FeTiO$_3$)
    Calcium ferrite type of structure (CaFeO$_2$)

Jadeite ($NaAlSi_2O_6$)    CN of Si = 4
Garnet $(Mg,Fe)_3Al_2Si_2O_6$

*Lower Mantle* (1000–2900 kbar)

Wüstite (FeO)
Periclase (MgO)
Stishovite ($SiO_2$)    CN Si = 6

These conclusions are summarized in Fig. 6.

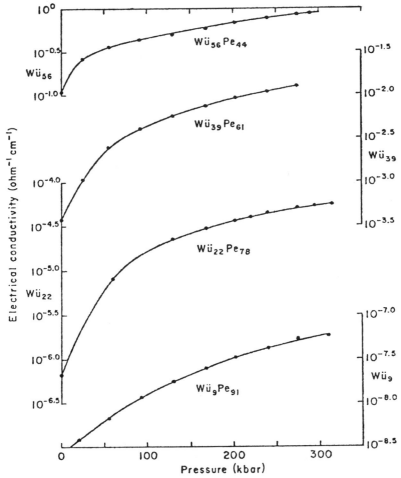

**Fig. 5** Electroconductivity of four magnesiowüstites as a function of pressure. [From Mao (1973c).]

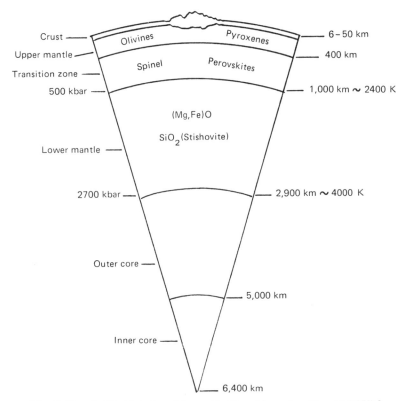

*Fig. 6.* The stratified interior of the earth. [Based in part on Bassett (1979).]

## 2. ELECTRICAL CONDUCTOR APPLICATIONS

Considerable interest has developed in the area of linear-chain or psuedo-one-dimensional compounds. By virtue of their crystalline packing, these materials exhibit anisotropy in certain physical variables such as magnetic, optical, and electrical properties. The electrical properties are of considerable interest, since they have generated a determined search for new and better electrical conductors.

We can define four broad categories of one-dimensional (1-D) compounds:

    (a)  transition metal complexes,
    (b)  organic donor–acceptor complexes such as tetrafulvalenium 7,7,8,8,-tetracyano-*p*-quinodimethanide (TTF–

TCNQ), (TMTSF)$_2$ClO$_4$, where TMTSF represents tetramethyltetraselenatetrafulvalene,
(c) polymeric systems, i.e., polysulfurnitride (SN)$_x$ and polyacetylene, and
(d) ion polymers, i.e., PEO · LiX, where PEO represents polyethylene oxide [see Papke *et al.* (1982)].

For reviews on some of these systems, see Miller and Epstein (1976), Williams and Schultz (1979), Stucky *et al.* (1977), Marks (1978), Ferraro (1982), and Street and Clarke (1981).

### a. 1-D Tetracyanato Complexes

These materials demonstrate electrical conductivity along the metal chain and are insulators in orthogonal directions. The structural basis for the interesting electrical conductivity in these compounds is the stacking of square-planar moieties. These Pt(II) compounds form columnar stacks containing infinite chains of Pt–Pt atoms. However, differences in Pt–Pt distances exist. Decreases in Pt–Pt distances cause an increase in electrical conductivity along the chain, and compounds are obtained that manifest metallic properties. This metallic behavior has been attributed to the electron delocalization occurring along the chain involving the overlapping Pt $5dz^2$ orbitals. Figure 7 illustrates the chains in K$_2$Pt(CN)$_4$, a poor conductor, and in K$_2$Pt(CN)$_4$Br$_{0.3}$(KCP), an excellent conductor. One can observe the shorter Pt–Pt distances in the latter compound.

The nature of these linearly stacked complexes makes them of interest from a standpoint of pressure sensitivity. One would expect that pressure should increase the $dz^2$ overlap, causing an increase in conductivity, at least initially. In compounds with smaller Pt–Pt distances pressure sensitivity would be expected to be less.

*(1) Krogmann Salts* Interrante and Bundy (1972) examined the Krogmann salt K$_2$Pt(CN)$_4$Br$_{0.3}$ · 2.3H$_2$O at pressures of up to 100 kbar. Using x-ray diffraction methods under pressure, they found that the a and c unit cell dimensions decreased with pressure. The a parameter changed by 0.47 Å in 70 kbar, while c axis (Pt–Pt chain axis) was less pressure sensitive and much less sensitive than the chain axis in Pt(NH$_3$)$_4$PtCl$_4$ [Magnus green salt (MGS)] (Interrante and Bundy, 1971). This presumably is due to the shorter distance in the Krogmann salt (2.89 Å) as compared to 3.25 Å in MGS, as well

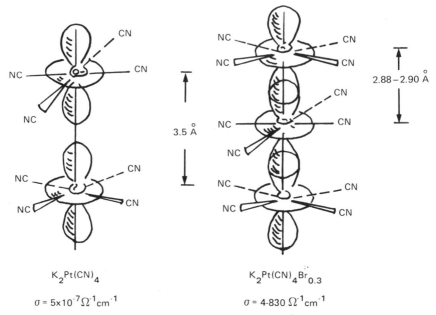

**Fig. 7.** Comparison of chains of square-planar Pt(CN)$_4^{x-}$ groups in K$_2$Pt(CN)$_4$ and K$_2$Pt(CN)$_4$Br$_{0.3}$.

as to the difference in charge on the metal atoms. Alternating current conductivities showed a slight increase (factor of 4) with a maximum at ~25 kbar, with a steady decrease thereafter due to the onset of steric and repulsive effects.

A metal–semiconductor phase transition in K$_2$Pt(CN)$_4$Br$_{0.3}$ · 3H$_2$O at ~20 kbar was observed (Müller and Jerome, 1974; Thielemans *et al.*, 1976), and its pressure dependency followed (Thielemans *et al.*, 1976).

A difference in pressure response was noted between a cation- and an anion-deficient tetracyanoplatinate (Hara *et al.*, 1975). The cation-deficient salt showed a slight drift of resistance upward with pressure, whereas the anion-deficient compound increased in resistance by a factor of 100 between 40 and 100 kbar after displaying a minimum at ~25 kbar (Interrante and Bundy, 1972).

*(2) Other Linear 1-D Complexes* Interrante and Browall (1974) conducted a high-pressure study of Pd and Pt complexes of the type M(A)X$_3$, where A represents (NH$_3$)$_2$, (C$_2$H$_5$NH$_2$)$_4$, en, etc.,

## 2. Electrical Conductor Applications

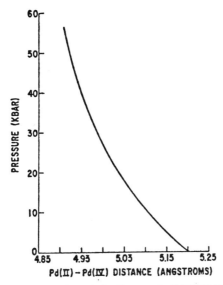

*Fig. 8.* The Pd(II)–Pd(IV) intrachain distance in Pd(NH$_3$)$_2$Cl$_3$ as a function of pressure. [Reprinted with permission from Interrante *et al.* (1974). Copyright 1974 American Chemical Society.]

and X represents halogen. These compounds show square-planar coordinated M(II) ions and octahedral coordinated M(IV) ions alternately arranged to give linear chains of

$$(---M(II)----X—M(IV)—X----)_n$$

along one crystallographic axis. Although no first-order phase changes occurred as determined by x-ray diffraction measured under pressure, an overall decrease in cell dimensions was observed. Figure 8 shows the Pd(II)–Pt(IV) distance versus pressure. Continuous and reversible increases in conductivity occurred by 9 orders at ~140 kbar. Above 100 kbar a gradual leveling off in conductivity was seen. Figure 9 shows the conductivity versus pressure plot for several of these complexes. Similar studies were conducted by Interrante and Bundy (1977) on mixed valence compounds Pt–(NH$_3$)$_2$Br$_3$ and Au(DBS)X$_2$, where DBS represents dibenzylsulfide. An enormous enhancement in conductivity occurred with an increase in pressure. The increase in conductivity in these complexes was attributed to a compression along the 1-D chains.

Onodera *et al.* (1979) applied pressure to platinum(II) glyoximates. Each Pt atom is coordinated to four nitrogen atoms of the

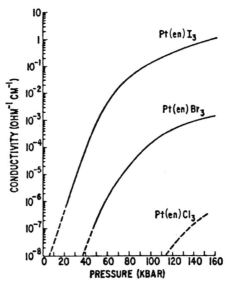

**Fig. 9.** Conductivity of the Pt(en)X$_3$ complexes as a function of pressure. [Reprinted with permission from Interrante et al. (1974). Copyright 1974 American Chemical Society.]

two dimethylglyoxime anions. Planar molecules of the complexes are arranged in parallel but are staggered to each other with Pt–Pt distances of about 3.23 Å. Resistivities of $10^{15}$ Ω cm are decreased significantly with pressure [$10^{16}$ at 40 kbar for Pt(DMG$_2$)] before they level off.

Onodera et al. (1975) examined several conductive metal phthalocyanines under very high pressures and found an appreciable increase in conductivity. Results are tabulated in Table V. Vanadyl phthalocyanine manifested the least resistivity, and each conductor showed a minimum in resistivity.

Hara et al. (1975) measured the pressure sensitivity of four Pt complexes such as SrPt(CN)$_4$ · 2H$_2$O, MgPt(CN)$_4$ · 7H$_2$O, BaPt(CN)$_4$ · 4H$_2$O, and Pt(NH$_3$)$_4$Pt(CN)$_4$. He failed to observe a resistance minimum for these materials as was observed for other d$^8$ complexes, and he attributed this to a considerable contribution of the $\pi^*$ orbitals of the CN$^-$ ligand to the conduction.

Thus far, experimental results concerning effects of pressure appear to substantiate predicted expectations for 1-D tetracyanates.

## 2. Electrical Conductor Applications

**Table V**

*Pressure Effects on Resistivities for Several Conductive Metal Phthalocyanines*

| Pc | Resistivity ($\Omega$ cm) | Pressure (kbar) |
|---|---|---|
| VOPc | $1 \times 10^{-1}$ | 290 |
| ThPc | $2 \times 10^{-1}$ | 330 |
| SnPc$_2$ | $2 \times 10$ | 360 |
| PbPc | 2 | — |
| CuPc | 4 | 620 |
| ZnPc | $1 \times 10$ | 640 |
| H$_2$Pc | $4 \times 10$ | 570 |
| CuPcCl$_{16}$ | $8 \times 10^{15}$ | 520 |

Because Pt–Pt distances are short in the Krogmann salts, minimal pressure responses of conductivity are observed. Although predictions (Thielemans *et al.*, 1976) have been made that at $P > 70$ kbar, KCP might become a superconductor at Tc 1–6 K; no such transition has been reported to date.

It should be noted that although most of the work reported heretofore has been made with a DAC, some work has been done with other pressure devices.

### b. Conducting Polymers

For the conducting polymers two general types can be distinguished: (1) conducting organic polymers (i.e., polyacetylene) and (2) polysulfurnitride (SN)$_x$. These materials have generated considerable interest because they exhibit metallic and even superconducting properties.

*(1) Organic Conducting Polymers*  Polyacetylene (CH)$_x$ was the first covalent organic polymer developed (Shirakawa *et al.*, 1977; MacDiarmid and Heeger, 1980). By means of chemical doping, the material could be transformed from a semiconductor to a metal. Polyacetylene (CH)$_x$ is the simplest conjugated organic polymer. It exists in two isomeric forms or as a mixture of the two forms, depending on the temperature at which the synthesis is made. The structures of the forms *cis*-(CH)$_x$ and *trans*-(CH)$_x$ are shown.

cis

trans

Doping can occur in several ways. $p$-Type doping involves the oxidation of $(CH)_x$ (*cis* or *trans*) films by exposure to $Br_2$, $I_2$, $AsF_5$, $HClO_4$, $H_2SO_4$, $(NO_2)(PF_6)$, etc. (MacDiarmid and Heeger, 1980). $p$-Type doping can be accomplished by electrochemical means by which $(ClO_4)^-$ or $PF_6^-$ anions are introduced (MacInnes *et al.*, 1981). $n$-Type doping involves the reduction of $(CH)_x$ and is accomplished by immersing the $(CH)_x$ films in a solution of lithium or potassium naphthalide in tetrahydrofuran solution (MacDiarmid and Heeger, 1980; Clarke and Street, 1980). $n$-Type doping can also be conveniently accomplished by electrochemical means (MacInnes, 1981). Doping of polyacetylene increases the conductivity from $10^{-9}$ $(\Omega\ cm)^{-1}$ to $10^2\ (\Omega\ cm)^{-1}$.

Other organic polymers that are found to be conductive can be synthesized. Diaconu *et al.* (1979) prepared poly(phenylacetylene), Pochan *et al.* (1980) synthesized poly(1,6-heptadiyne), Wnek *et al.* (1979) synthesized poly($p$-phenylene vinylenes), and Rabolt *et al.* (1980) and Chance *et al.* (1980) prepared poly($p$-phenylene sulfide). Polypyrrole was synthesized by Kanazawa *et al.* (1979) and Diaz *et al.* (1979), and poly($p$-phenylene) by Ivory *et al.* (1979).

The conductivity relationships for these organic polymers versus conductivities of classical conductors are illustrated in Fig. 10.

No extensive pressure investigation have been made on these polymers. Seeger *et al.* (1978) studied $AsF_5$-doped $(CH)_x$ at pressures to 10 kbar and found that conductivity was not affected significantly. Doped phthalocyanines (PcAlF, PcGaF) (Webb *et al.*, 1982) have been examined at pressures of up to 65 kbar. These materials stack and after partial oxidation demonstrate appreciable electrical conductivities. It would be expected that pressures would vary the interstacking distances. In general, resistances were found to drop by one or two orders of magnitude. It would appear that more pressure studies of these materials, combined with other diagnostic methods such as vibrational spectroscopy, are in order.

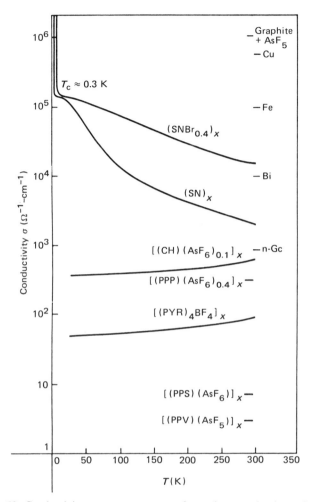

**Fig. 10.** Conductivity versus temperature for various conducting polymers and classical conductors: PPP = poly(phenylene), PYR = pyrrole, PPS = poly(p-phenylene sulfide), PPV = poly(p-phenylene vinylene). [From Street and Clarke (1981). Copyright 1981 by International Business Machines Corporation; reprinted with permission.]

*(2) Polymeric* (SN)$_x$   Walatka *et al.* (1973) found that polymeric (SN)$_x$ crystals are metallic, and Mikulski *et al.* (1975) found that the substances possess a zigzag structure of (SN)$_x$ chains. Greene *et al.* (1975) found that the material became superconducting at <0.3 K. Few pressure studies have been made on the materials to date. Gill *et al.* (1975) found that the conductivity of the material increased with pressure to 9 kbar and subsequently decreased. The critical temperature $T_c$ was found to increase slightly with pressure to 0.6 K. Friend *et al.* (1977) measured the pressure dependency of conductivity for (SN)$_x$ up to 40 kbar.

Unfortunately, little information is available concerning the structure of these chemically conducting polymers to enable one to make any reasonable predictions as to the effects of pressure. For brominated (SN)$_x$ it has been shown that the reduced form of the dopant intercalates between polymeric cation chains of the oxidized polymer (Street and Gill, 1979). Figure 11 shows the orientation of Br$_3^-$ ions relative to the (SN)$_x$ chains. For (CH)$_x$ the dopant enters the lattice and causes an expansion of the (100) plane spacing (Baughman *et al.*, 1978; Clarke and Street, 1980). Table VI presents the d$_{(100)}$ and d$_{(200)}$ spacings for these materials. Figure 12 illustrates the crystal structure of *cis*(CH)$_x$. The structures of other organic conducting polymers are unavailable to date.

### c. Organic Charge-Transfer Salts

Typical of this class of materials is TTF–TCNQ. Salts of this type demonstrate TTF radical cations and TCNQ radical anions in

*Table VI*

*Interplanar Spacing of* (CH)$_x$ *Films after Intercalation*

| Film | $d_{(100)}$ (Å) | $d_{(200)}$ (Å) | $D$ (Å)[a] |
|---|---|---|---|
| (CH)$_x$ | — | 3.80 | — |
| (CHBr$_{0.47}$)$_x$ | 7.31 | 3.63 | 3.46 |
| (CHI$_{0.20}$)$_x$ | 7.96 | 3.93 | 4.06 |
| [CH(AsF$_5$)$_{0.16}$]$_x$ | 8.83 | 4.39 | 4.98 |

[a] Van der Waals diameter $D$ as determined by $D$ (Å) = $2d_{(200)}$ − 3.80, where 3.80 Å is diameter of a (CH)$_x$ chain.

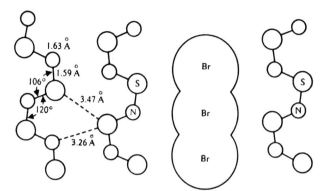

**Fig. 11.** Structure of brominated $(SN)_x$ showing the orientation of $Br_3^-$ ions relative to the $(SN)_x$ chains. [From Street and Clarke (1981). Copyright 1981 by International Business Machines Corporation; reprinted with permission.]

columnar stacks with interplanar spacings of 3.47 and 3.17 Å, respectively, and these distances are shorter than distances calculated from the usual van der Waals distances. Pressure effects would be expected to shorten these distances and to increase the overlap of the $\pi$-electron clouds; this would result in an increase in conductivity. Also of interest is the effect of pressure on $T_c$, the temperature

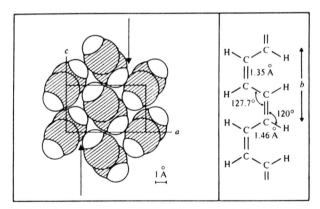

**Fig. 12.** Crystal structure of $cis$-$(CH)_x$. The arrows indicate how intercalation takes place between planes of $(CH)_x$ chains; crystal class: orthorhombic; space group: Pnma, where $a = 7.61$ Å, $b = 4.47$ Å, and $c = 4.39$ Å. [From Street and Clarke (1981). Copyright 1981 by International Business Machines Corporation; reprinted with permission.]

of the metal–insulator transition. Salts of the TCNQ anion radical have been studied under pressure. Aust et al. (1964) studied $((C_2H_5)_3NH^+)TCNQ_2^-$, (quinolinium$^+$)TCNQ$_2^-$, and KTCNQ salts at pressures of 140–150 kbar. A decrease in resistance was observed, followed by a large rise with drift. The changes observed were not reversible. Infrared studies indicated that reaction took place between adjacent TCNQ moieties through the C ≡ N groups. Debray et al. (1977) and Chu et al. (1973) studied TTF–TCNQ under pressure. The latter studies were done at 20 kbar and at 20–273 K. The metal–insulator $T_c$ increased 1 K kbar$^{-1}$. The conductivity increased by a factor of 4.5 at 20 kbar. Shirotani et al. (1975) examined alkali metal salts of TCNQ. Resistances dropped significantly up to ~150 kbar. Above this pressure, rises in resistances took place. Onodera et al. (1979) studied TTF–TCNQ under pressure and found that the resistivity showed a minimum at ~43 kbar. Thereafter, the resistance drifts upward when the pressure is kept constant. On decreasing the pressure, the resistance rises. The series of changes observed is irreversible. As in previously discussed studies, the occurrence of a chemical reaction in the solid state under pressure is a possibility. Salts of the TTF cation radical containing halogen anions were also studied. The resistance of the salts TTF–Br$_{0.8}$ and TTF–I$_5$ exhibit an irreversible upward drift from the minimum, again suggesting a chemical reaction occurring under pressure. DMTTF–TCNQ and TMTTF–TCNQ salts were also compressed. Results paralleled those of TTF–TCNQ, indicating that methyl groups had no effect on conductivity. Table VII compiles the minimum resistivity observed for several compounds of the charge-transfer type.

Another group of charge-transfer quasi-one-dimensional organic conductors belongs to the series (TMTSF)$_2$X (where TMTSF is tetramethyltetraselenafulvalene and X is an inorganic counterion such as ClO$_4$, PF$_6$, AsF$_6$, TaF$_6$, SbF$_6$, or ReO$_4$. These are called the Beckgaard salts. Jerome (1981) found that (TMTSF)$_2$PF$_6$ becomes superconductive at pressures <11 kbar and at 4.2 K. For example, the longitudinal conductivity exceeds $10^5$ ($\Omega$ cm)$^{-1}$. Parkin et al. (1981) reported that (TMTSF)$_2$SbF$_6$ and (TMTSF)$_2$TaF$_6$ become superconductive near 1 K when sufficient pressure is applied (>10 and 11 kbar, respectively). They also found that (TMTSF)$_2$ClO$_4$ becomes superconductive at pressures of <3 kbar and that at 1.5 kbar there is a minimum in p above the superconducting phase transition temperature. New data show that (TMTSF)$_2$ClO$_4$ is superconduct-

*Table VII*

*Minimum Resistivity for Several Charge-Transfer Compounds*

| Materials | $\rho_{min}$ ($\Omega$ cm) | Pressure (kbar) |
|---|---|---|
| TTF–TCNQ | $8.2 \times 10^{-3}$ | $43.0 \pm 3.5$ |
| DMTTF–TCNQ | $7.2 \times 10^{-3}$ | $43.0 \pm 3.5$ |
| TMTTF–TCNQ | $7.0 \times 10^{-3}$ | $43.0 \pm 3.5$ |
| TTF–Br$_{0.8}$ | $1.0 \times 10^{-1}$ | $18.0 \pm 1.0$ |
| TTF$_7$–I$_5$ | $4.4 \times 10^{-2}$ | $41.0 \pm 3.0$ |
| TTF | $6.0 \times 10$ | 250 |
| DMTTF | $9.0 \times 10^5$ | 250 |
| TMTTF | $5.0 \times 10^2$ | 250 |

ing at 1.2 K and at ambient pressure (see Bechgaard *et al.*, 1982). These are the first organic superconductors found to date (Bechgaard and Jerome, 1982). The recent research on the Bechgaard salts demonstrates the importance of applying pressure to conductive materials.

## 3. LUBRICANT STUDIES

The application of high-pressure interferometry in the study of lubricants is an unusual practical application for the DAC. Since, in this special case, only one diamond window is used, it has been necessary to use emission IR or reflection methods. Figure 13 shows the apparatus used by Lauer and Peterkin (1975a) and Lauer (1978) to study lubricants. The technique allows one to study lubricants under conditions that prevail in a bearing contact. Vibrational spectroscopy can determine whether changes such as phase changes, glass transitions, conformational changes, and chemical changes occur under pressure. Figure 14 shows an emission spectrum of a bearing fluid consisting of hydrogenated monosubstituted aromatics and straight-chain hydrocarbons and compares it with an absorption spectrum of the fluid (Lauer and Peterkin, 1975a,c). A 1 : 1 correlation is noted. The emission spectrum under pressure, however, shows changes in intensities at 700 and 750 cm$^{-1}$. Lauer and Peterkin (1975b) have studied other fluids under dynamic conditions by using this technique and have used vibrational spectroscopy to identify any changes occurring.

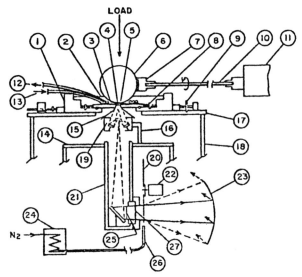

*Fig. 13.* Sliding bearing contact and interferometer attachment for emission measurements: (1) fluid cup, (2) fluid, (3) diamond holder, (4) diamond window, (5) contact region, (6) rotating ball, (7) holder, (8) diamond holder $x/y$ adjustment screw, (9) fluid cup $x/y$ adjustment screw, (10) flexible shaft, (11) motor drive, (12) cooling water out, (13) cooling water in, (14) interferometer source compartment cover, (15) aluminum reflector, (16) Beck lens support, (17) bearing table surface, (18) bearing table support, (19) Beck lens, (20) chopper blade, (21) adapter tube for emission spectra, (22) chopper motor, (23) interferometer collimating mirror, (24) gas cooler, (25) real image of contact region, (26) gas nozzle, and (27) infrared transparent window. [From Lauer (1978).]

## 4. FORENSIC SCIENCE

### a. Introduction

The versatility of the diamond anvil cell has been emphasized throughout this book. The ability to obtain infrared spectra of microsamples is an important feature of the DAC. This capability has been cited by Lippincott *et al.* (1961), Krishnan *et al.* (1980) and Ferraro and Basile (1979). The method is rapid and nondestructive and has lent itself to the examination of evidential materials, such as fibers, hair, paints, plastics, explosives, and others. The technique has proven to be extremely useful and has established itself as a powerful tool in the forensic laboratory.

## 4. Forensic Science

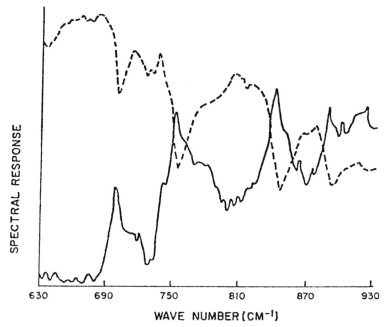

*Fig. 14.* Infrared absorption (broken curve) and emission (solid curve) of a bearing fluid at 160°C and 5 kbar. [From Lauer (1978).]

The first report of the use of the DAC for possible forensic purposes was made by Eves (1963) at the First Interpol Seminar on the Scientific Aspects of Police Work (Paris, France, 1963). Corrigan (1965, 1966) later cited the use of the technique in forensic work. A review of the technique by Midkeff *et al.* appeared in 1979.

This section attempts to present an overview of the use of the DAC in forensic science.

### b. Experimental Technique

The DAC has been described extensively in Chapter 2. As we learned previously, the usable regions of the infrared spectrum with the DAC are 400–2700 cm$^{-1}$ and 1800–50 cm$^{-1}$. The usable range of sapphire is 400–1700 cm$^{-1}$ and the far-infrared region. See Chapter 2 for more details and Figs. 7 and 8 therein for spectra of diamonds and sapphires.

The advantages of the DAC for forensic work are the following:

(a) It is a microtechnique (1–5 $\mu$g).
(b) No sample preparation is necessary.
(c) It is a nondestructive technique.
(d) The method is rapid and reliable.

## c. Results

The results recorded for a number of substances of interest in forensic microanalysis are presented. It is not our intention to present all data obtained with the DAC but to give the reader some idea of how useful the technique can be. One interesting area has been the vehicle collision and/or hit-and-run case, in which the automobile paint becomes an important piece of evidence. From the paint identification, the type of vehicle involved can be ascertained. Usually, only submicrogram quantities of paint are available, and the DAC can very readily produce an adequate infrared spectrum. Extensive studies of paints and their binders, which include alkyd and acrylic melamine enamels, polyurethane enamels, epoxies, acrylic lacquers, and nitrocellulose, have been made (see Tweed *et al.*, 1974; Rodgers *et al.*, 1976a–c; Cartwright and Rodgers, 1976). Identifications of pigments and fillers such as talc, silicon, and clay have also been made. Extensive studies on topcoat and undercoat paints and body fillers were made for a number of originally painted and repainted automobiles available in Canada. These spectra have been used to create a reference file for paints and vehicle identification (Cartwright *et al.*, 1976). Rodgers *et al.* (1976a) list a series of case histories in which paint identification has been important. Typical paint spectra obtained in a DAC are illustrated in Figs. 15 and 16. Figure 17 shows a spectrum of an acrylic enamel paint chip.

The DAC has been instrumental in obtaining spectra of other materials, which have aided in the solution of various crimes. Examples of these involve fibers from rugs, clothing, curtains, furniture, and hair from animal and human sources. Figure 18 presents a DAC spectrum of a polyester fiber (Read and Kopec, 1978). Figure 19 shows a spectrum of an acrylic rug yarn. Figure 20 presents spectra of human hair.

Other substances that have been examined in the DAC include explosives. In investigations of criminal bombings, fragments and explosive particles are important pieces of evidence and can be

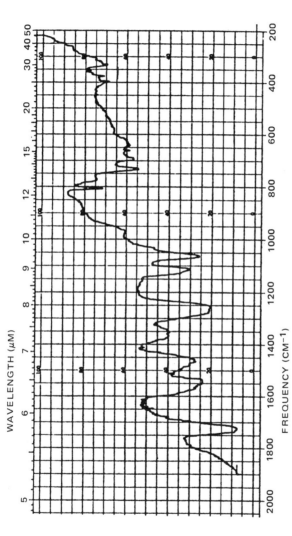

*Fig. 15.* DAC spectrum of a blue alkyd melamine topcoat (1972 Toyota Celica) (P-E 621). [From Tweed *et al.* (1974).]

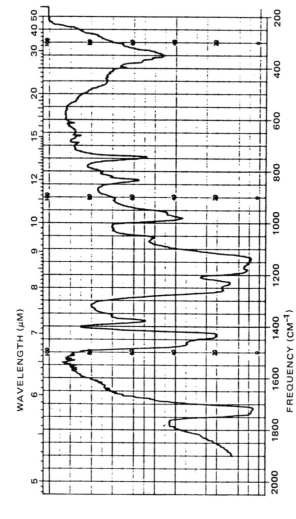

**Fig. 16.** DAC spectrum of methylmethacrylate (from a taillight lens) (P-E 621). [From Tweed *et al.* (1974).]

## 4. Forensic Science

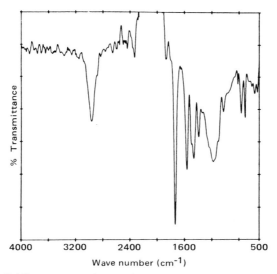

**Fig. 17.** DAC spectrum of water-based acrylic enamel paint chip. [From Krishnan and Ferraro (1982).]

identified by the use of the DAC (see Kopec *et al.*, 1978; Washington and Midkeff, 1976; Washington *et al.*, 1977a,b; Midkeff *et al.*, 1979). Figure 21 illustrates a spectrum obtained in a DAC using sapphire rather than diamond of high-explosive RDX (cyclotrimethylenetrinitramine) containing 5% HMX (cyclotetramethylenetetranitramine).

Gunshot residue from the hands of a person suspected of firing a gun has also been identified with the help of the DAC (Goleb and Midkeff, 1975; Kopec *et al.*, 1984).

Additional uses for the DAC have surfaced owing to the forensic interest in adhesives (Kopec *et al.*, 1978; Read and Beckman, 1982, unpublished data).

The use of the technique in the identification of drugs and narcotics has been reported by Humphrey (1978).

Some examples have been presented to illustrate that the DAC has become an important and powerful tool in the forensic laboratory. It is anticipated that additional uses will be found in forensic science and in other areas in the future.

A rather unique use of the DAC has recently been reported. Materials of interest in art and archaeology have been analyzed with the DAC (Laver and Williams, 1978).

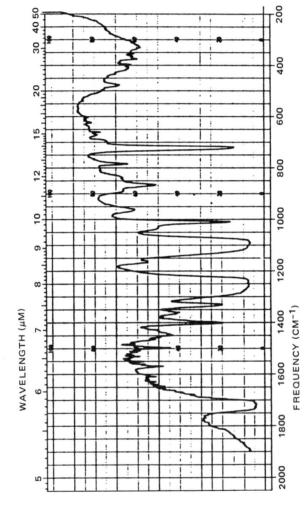

*Fig. 18.* DAC spectrum of a polyester fibre (Millhaven Fortrel) (P-E 621). [From Tweed *et al.* (1974).]

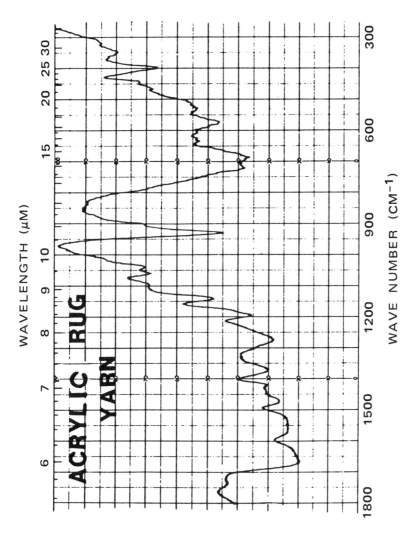

*Fig. 19.* Spectrum of an acrylic rug yarn. [Reprinted from Ferraro and Basile, *American Laboratory*, Volume 11, Number 3, page 31, 1979. Copyright 1979 by International Scientific Communications, Inc.]

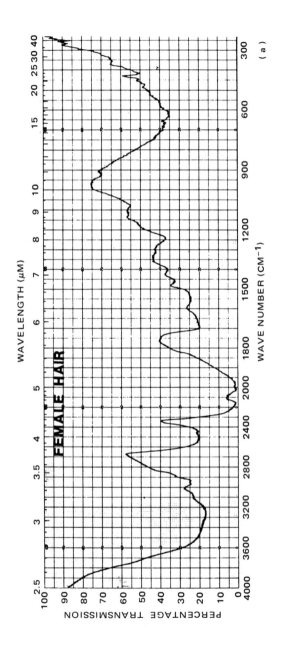

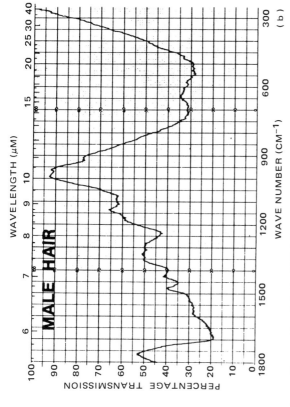

*Fig. 20.* (a) DAC spectrum of human female hair, (b) DAC spectrum of human male hair (black—no gray). [Reprinted from Ferraro and Basile, *American Laboratory*, Volume 11, Number 3, page 31, 1979. Copyright 1979 by International Scientific Communications, Inc.]

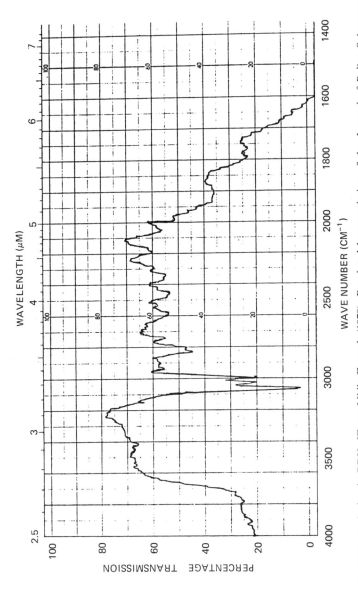

**Fig. 21.** HMX explosive in RDX. [From Midkeff et al. (1979). Reprinted by permission of the *Journal of Police Science and Administration*, copyright 1979 by the International Association of Chiefs of Police, Inc., Vol. 7, No. 4, p. 433.]

## REFERENCES

*Geochemistry and Geophysics*

Ahrens, T. J. (1980). *Science* **207**, 1035.
Altschuler, L. V., and Kormer, S. B. (1961). *Bull. Acad. Sci. U.S.S.R., Geophys. Ser.* **1**, 18.
Barnett, J. D., Block, S., and Piermarini, G. J. (1973). *Rev. Sci. Instrum.* **74**, 1.
Bassett, W. A. (1979). *Annu. Rev. Earth Planet. Sci.* **7**, 357.
Bassett, W. A., and Brody, E. M. (1977). "High Pressure Research," pp. 519–531. Academic Press, New York.
Bassett, W. A., and Takahashi, T. (1974). *Adv. High Pressure Res.* **4**, 165.
Bassett, W. A., Takahashi, T., and Stook, P. W. (1967). *Rev. Sci. Instrum.* **38**, 37.
Bassett, W. A., Wilburn, D. R., Hrubec, J. A., and Brody, E. M. (1979). *High Pressure Sci. Technol. AIRAPT Conf., 1977*, **2**, 75–84.
Bell, P. M. (1983). *Revs. Geophys. Space Phys.* **21**, 1394.
Bell, P. M., and Mao, H. K. (1975). *Annu. Rev. Geophys. Lab., Washington, D.C.* **74**, 400.
Bell, P. M., Mao, H. K., Weeks, R. A., and Van Valkenburg, A. (1976). *Annu. Rev. Geophys. Lab., Washington, D.C.* **75**, 515.
Birch, F. (1952). *J. Geophys. Res.* **57**, 227.
Birch, F. (1960). *J. Geophys. Res.* **65**, 1083.
Birch, F. (1961). *J. Geophys. Res.* **66**, 2199.
Block, S., and Piermarini, G. J. (1976). *Phys. Today* **29**, 44.
Drickamer, H. G., and Frank, C. W. (1973). "Electronic Transitions and the High Pressure Chemistry and Physics of Solids." Chapman and Hall, London.
Hazen, R. M. (1977a). *EOS Trans. Am. Geophys. Union* **58**, 518.
Hazen, R. M. (1977b). *Phys. Chem. Miner.* **1**, 83.
Hazen, R. M., and Finger, L. W. (1978). *Science* **201**, 1122.
Hazen, R. M., and Prewitt, C. T. (1977). *Am. Mineral.* **62**, 309.
Huggins, F. E., Mao, H. K., and Virgo, D. (1975). *Annu. Rep. Geophys. Lab., Washington, D.C.* **74**, 405.
Hughes, D. S., and McQueen, R. M. (1958). *Trans. Am. Geophys. Union* **39**, 959.
Liu, L.-G. (1975a). *Earth Planet. Sci. Lett.* **24**, 357.
Liu, L.-G. (1975b). *Earth Planet. Sci. Lett.* **25**, 286.
Liu, L.-G. (1976). *Earth Planet. Sci. Lett.* **33**, 101.
Liu, L.-G. (1978). *Science* **199**, 422.
Lombard, D. B. (1961). Univ. of California Res. Lab. Rep. No. 6311.
McQueen, R. G., and Marsh, S. P. (1966). *Mem. Geol. Soc. Am.* **97**, 153.
Mao, H. K. (1973a). *Annu. Rep., Geophys. Lab., Washington, D.C.* **72**, 552.
Mao, H. K. (1973b). *Annu. Rep., Geophys. Lab., Washington, D.C.* **72**, 544.
Mao, H. K. (1973c). *Annu. Rep., Geophys. Lab., Washington, D.C.* **72**, 557.
Mao, H. K. (1974). *Annu. Rep. Geophys. Lab., Washington, D.C.* **73**, 510.
Mao, H. K. (1976). *Phys. Chem. Miner. Rocks NATO Adv. Study, Inst.* "Petrophys," p. 573.
Mao, H. K., and Bell, P. M. (1971). *Annu. Rep. Geophys. Lab., Washington, D.C.* **70**, 176.

Mao, H. K., and Bell, P. M. (1972a). *Annu. Rep. Geophys. Lab., Washington, D.C.* **71,** 520.
Mao, H. K., and Bell, P. M. (1972b). *Annu. Rep. Geophys. Lab., Washington, D.C.* **71,** 524.
Mao, H. K., and Bell, P. M. (1972c). *Annu. Rep. Geophys. Lab., Washington, D.C.* **71,** 527.
Mao, H. K., and Bell, P. M. (1975). *Annu. Rep. Geophys. Lab., Washington, D.C.* **73,** 209.
Mao, H. K., and Bell, P. M. (1977). *High Pressure Res. Appl. Geophys. Pap. U.S. Jpn. Jt. Semin. 1976,* pp. 493–502.
Mao, H. K., and Bell, P. M. (1978). *Science* **200,** 1145.
Mao, H. K., and Bell, P. M. (1981). *Rev. Sci. Instrum.* **52,** 615.
Mao, H. K., Virgo, D., and Bell, P. M. (1977). *Annu. Rep. Geophys. Lab., Washington, D.C.* **76,** 522.
Mao, H. K., Bell, P. M., Shaner, J. W., and Steinberg, D. J. (1978). *J. Appl. Opt.* **49,** 3276.
Merrill, L., and Bassett, W. A. (1974). *Rev. Sci. Instrum.* **45,** 290.
Ming, L.-C., and Bassett, W. A. (1975). *Science* **187,** 66.
Piermarini, G. J., and Block, S. (1975). *Rev. Sci. Instrum.* **46,** 973.
Piermarini, G. J., and Weir, C. E. (1962). *J. Res. Natl. Bur. Stand. Sect. A* **66,** 325.
Ringwood, A. E. (1962). *Geochim. Cosmochim. Acta* **26,** 457.
Ringwood, A. E. (1963). *Nature (London)* **198,** 79.
Ringwood, A. E. (1966). *Earth Planet. Sci. Lett.* **1,** 241.
Ringwood, A. A. (1975). "Composition and Petrology of the Earth's Mantle," p. 188. McGraw-Hill, New York.
Ringwood, A. E., and Major, A. (1966). *Earth Planet. Sci. Lett.* **1,** 135.
Ruoff, A. L. (1979). *Cornell Q. Eng.* **14,** 2.
Sharma, S. K., Mao, H. K., and Bell, P. M. (1980). *Phys. Rev. Lett.* **44,** 886.
Wackerle, J. (1962). *J. Appl. Phys.* **33,** 922.
Walling, P. L., and Ferraro, J. R. (1978). *Rev. Sci. Instrum.* **49,** 1557.
Weir, C. E., Piermarini, G. J., and Block, C. (1969). *Rev. Sci. Instrum.* **40,** 1133.
Whitfield, C. H., Brody, E. M., and Bassett, W. A. (1976). *Rev. Sci. Instrum.* **47,** 942.
Yagi, T., and Akimoto, S. (1976). *Tectonophysics* **35,** 259.

*Electrochemical Conductors*

Aust, R. B., Samara, G. A., and Drickamer, H. G. (1964). *J. Chem. Phys.* **41,** 2003.
Baughman, R. H., Hsu, S. L., Pez, G. P., and Signorelli, A. J. (1978). *J. Chem. Phys.* **68,** 5405.
Bechgaard, K., and Jerome, D. (1982). *Sci. Amer.* **247**(1), 52.
Bechgaard, K., Carneiro, K., Eg, O., Olsen, M., Rasmussen, F. B., Jacobsen, C., and Rindorf, G. (1982). *Mol. Cryst. Liq. Cryst.* **79,** 271.
Chance, R. R., Shacklette, L. W., Miller, G. G., Ivory, D. M., Sowa, J. M., Elsenbaumer, R. L., and Baughman, R. H. (1980). *J. Chem. Soc. Chem. Commun.* 348.
Chu, C. W., Harper, J. M. E., Geballe, T. H., and Greene, R. L. (1973). *Phys. Rev. Lett.* **31,** 1491.

# References

Clarke, T. C., and Street, G. B. (1980). *Synthetic Metals* **1**, 119.
Debray, D., Millet, R., Jerome, D., Barisic, S., Giral, L., and Fabre, J. M. (1977). *J. Phys. Lett. Orsay, Fr.* **38**, L227.
Diaconu, I., Dumitrescu, S., and Simionescu, C. (1979). *Environ. Polym. J.* **15**, 1155.
Diaz, A. F., Kanazawa, K. K., and Gardini, G. P. (1979). *J. Chem. Soc. Chem. Commun.* 635.
Ferraro, J. R. (1982). *Coord. Chem. Rev.* **43**, 205.
Friend, R. H., Jerome, D., Rehmatullah, S., and Yoffe, A. D. (1977). *J. Phys. C* **10**, 1001.
Gill, W. D., Greene, R. L., Street, G. B., and Little, W. A. (1975). *Phys. Rev. Lett.* **35**, 1732.
Greene, R. L., Street, G. B., and Suter, L. J. (1975). *Phys. Rev. Lett.* **34**, 577.
Hara, Y., Shirotani, I., and Onodera, A. (1975). *Solid State Commun.* **17**, 827.
Interrante, L. V., and Browall, K. W. (1974). *Inorg. Chem.* **13**, 1162.
Interrante, L. V., and Bundy, F. P. (1971). *Inorg. Chem.* **10**, 1169.
Interrante, L. V., and Bundy, F. P. (1972). *Solid State Commun.* **11**, 1641.
Interrante, L. V., and Bundy, F. P. (1977). *J. Inorg. Nucl. Chem.* **39**, 1333.
Interrante, L. V., Browall, K. W., and Bundy, F. P. (1974). *Inorg. Chem.* **13**, 1158.
Ivory, D. M., Miller, G. G., Sowa, J. M., Shacklette, L. W., Chance, R. R., and Baughman, R. H. (1979). *J. Chem. Phys.* **71**, 1506.
Jerome, D. (1981). *Int. Conf. Low-Dimens. Conduct., Boulder, Colo. 1981*.
Kanazawa, K. K., Diaz, A. F., Geiss, R. H., Gill, W. D., Kwak, J. F., Logan, J. A., Rabolt, J. F., and Street, G. B. (1979). *J. Chem. Soc. Chem. Commun.* 854.
MacDiarmid, A. G., and Heeger, A. J. (1980). *Synth. Met.* **1**, 101.
MacInnes, D., Druy, M. A., Nigrey, P. J., Nairns, D. P., MacDiarmid, A. G., and Heeger, A. J. (1981). *J. Chem. Soc. Chem. Commun.* 317.
Marks, T. J. (1978). *Ann. N.Y. Acad. Sci.* **313**, 594.
Mikulski, C. M., Russo, P. J., Saran, H. S., MacDiarmid, A. G., Garito, A. F., and Heeger, A. J. (1975). *J. Am. Chem. Soc.* **97**, 6358.
Miller, J. S., and Epstein, A. J. (1976). *Prog. Inorg. Chem.*, New York. **20**, 1–151.
Müller, W. H.-G., and Jerome, D. (1974). *J. Phys. Lett. Orsay, Fr.* **35**, L103.
Onodera, A., Kawaii, N., and Kobayashi, T. (1975). *Solid State Commun.* **17**, 775.
Onodera, A., Shirotani, I., Hara, Y., and Anzai, H. (1979). *High Pressure Sci. Technol. AIRAPT Conf., 1977* **1**, 498.
Papke, B. L., Ratner, M. A., and Shriver, D. F. (1982). *J. Electrochem. Soc.* **129**, 1434.
Parkin, S. S., Jerome, D., and Bechgaard, K. (1981). *Int. Conf. Low-Dimens. Conduct., Boulder, Colo., 1981*.
Pochan, J. M., Gibson, H. W., and Bailey, F. C. (1980). *J. Polym. Sci., Polym. Lett. Ed.* **18**, 447.
Rabolt, J. F., Clarke, T. C., Kanazawa, K. K., Reynolds, J. R., and Street, G. B. (1980). *J. Chem. Soc. Chem. Commun.* 347,
Seeger, K., Gill, W. D., Clarke, T. C., and Street, G. B. (1978). *Solid State Commun.* **28**, 873.
Shirakawa, H., Louis, E. J., MacDiarmid, A. G., Chang, C. K., and Meeger, A. J. (1977). *J. Chem. Soc. Chem. Commun.* 578.
Shirotani, I., Onodera, A., and Sakai, N. (1975). *Bull. Chem. Soc. Jpn.* **48**, 167.
Street, G. B., and Clarke, T. C. (1981). *IBM J. Res. Dev.* **25**, 51.
Street, G. B., and Gill, W. D. (1979). *NATO Conf. Ser., Ser. 6*, 301.

Stucky, G. D., Schultz, A. J., and Williams, J. M. (1977). *Annu. Rev. Mater. Sci.* **1,** 301.
Thielemans, M., Deltour, R., Jerome, D., and Cooper, J. R. (1976). *Solid State Commun.* **19,** 21.
Walatka, V. V., Labes, M. M., and Perlstein, J. M. (1973). *Phys. Rev. Lett.* **31,** 1139.
Webb, A. W., Brant, P., Nohn, R. S., and Weber, D. C. (1982). *Proc. Am. Chem. Soc. Meet. Las Vegas, Symp. Conduct. Polym., Div. Polym. Chem.*
Williams, J. M., and Schultz, A. J. (1979). *NATO Conf. Ser., Ser. 6,* 337–368.
Wnek, G. E., Chien, J. C. W., Karasz, F. E., and Lillya, C. P. (1979). *Polymer* **20,** 1441.

## Lubricant Studies

Lauer, J. L. (1978). *Fourier Transform Infrared Spectrosc.* **1,** 169–213.
Lauer, J. L., and Peterkin, M. E. (1975a). *Am. Lab. Fairfield, Conn.* **7,** 27.
Lauer, J. L., and Peterkin, M. E. (1975b). *Appl. Spectrosc.* **29,** 78.
Lauer, J. L., and Peterkin, M. E. (1975c). *J. Lubr. Technol.* **97,** 145.

## Forensic Science

Cartwright, N., and Rodgers, P. (1976). *J. Can. Soc. Forensic Sci.,* **21,** 862.
Cartwright, N., Cartwright, N. S., and Rodgers, P. G. (1976). *J. Can Soc. Forensic Sci.,* **10,** 7.
Corrigen, A. R. (1965). *Int. Criminology Congr., 5th, Montreal, 1965.*
Corrigen, A. R. (1966). *Conf. Chem. Inst. Can., 4th, Saskatoon, 1966.*
Eves, C. R. (1963). *Interpol Semin. Sci. Aspects Police Work, Toronto, 1963.*
Ferraro, J. R., and Basile, J. R. (1979). *Am. Lab. Fairfield, Conn.* **11,** 31.
Goleb, J., and Midkeff, C. R. (1975). *J. Forensic Sci.* **20,** 701.
Humphrey, M. (1978). *M.S. thesis, Northeastern Univ., Boston, Massachusetts.*
Kopec, R., Washington, W., and Midkeff, C. (1978). *J. Forensic Sci.* **23,** 57.
Kopec, R., Kinard, W., and Washington, W. (1984). In press.
Krishnan, K., and Ferraro, J. R. (1982). *In* "Fourier Transform Infrared Spectroscopy," (J. R. Ferraro and L. J. Basile, eds.) Vol. 3, pp. 149–209, Academic Press, New York.
Krishnan, K., Hill, S. L., and Brown, R. H. (1980). *Am. Lab. Fairfield, Conn.* **12,** 104.
Laver, M. E., and Williams, R. S. (1978). *JIIC—Canadian Group* **3,** 34.
Lippincott, E. R., Welsh, F., and Weir, C. (1961). *Anal. Chem.* **33,** 137.
Midkeff, C. R., Washington, W. D., and Kopec, R. J. (1979). *J. Police Sci. Admin.* **7,** 426.
Read, L. K., and Beckman, M. (1982). U.S. Postal Service, unpublished work.
Read, L. K., and Kopec, R. J. (1978). *J. Assoc. Off. Anal. Chem.* **61,** 526.
Rodgers, P. G., Cameron, R., Cartwright, N. S., Clark, W. H., Deak, J. S., and Norman, E. W. W. (1976a). *J. Can. Soc. Forensic Sci.* **9,** 103.
Rodgers, P. G., Cameron, R., Cartwright, N. S., Clark, W. H., Deak, J. S., and Norman, E. W. W. (1976b). *J. Can. Soc. Forensic Sci.* **9,** 49.

Rodgers, P. G., Cameron, R., Cartwright, N. S., Clark, W. H., Deak, J. S., and Norman, E. W. W. (1976c). *J. Can. Soc. Forensic Sci.* **9,** 1.
Tweed, F. T., Cameron, R., Deak, J. S., and Rodgers, P. G. (1974). *Forensic Sci.* **4,** 211.
Washington, W. D., and Midkeff, C. R. (1976). *J. Forensic Sci.* **21,** 862.
Washington, W. D., Midkeff, C. R., and Snow, K. (1977a). *J. Forensic Sci.* **22,** 329.
Washington, W. D., Kopec, R. J., and Midkeff, C. R. (1977b). *J. Assoc. Off. Anal. Chem.* **60,** 1331.

## BIBLIOGRAPHY

Ahrens, T. J., Jeanloz, R., and Mao, H. K. (1978). Discovery of a new phase of Calcia via shock-wave and diamond cell techniques and its implication for composition of lower mantle. *Trans. Am. Geophys. Union* **59,** 1181.
Bassett, W. A. (1979). The diamond cell and the nature of the earth's mantle. *Annu. Rev. Earth Planet. Sci.* **7,** 357.
Iye, Y., and Tanuma, S. (1982). Effect of pressure on superconductivity of graphite intercalation compounds $C_8KHg$, and $C_8RbHg$. *Solid State Commun.* **44,** 1.
Lacam, A., Madon, M., and Poirier, J. P. (1980). Olivine glass and spinel formed in a laser heated diamond-anvil high pressure cell. *Nature* **288,** 5787.
Liu, L.-G. (1978). A fluorite isotype of $SnO_2$ and a new modification of $TiO_2$: Implications for the earth's lower mantle. *Science* **199,** 422.
Liu, L.-G., and Ringwood, A. E. (1975). Synthesis of a perovskite-type polymorph of $CaSaO_3$. *Earth Planet. Sci. Lett.* **28,** 209.
Ming, L.-C., and Manghnani, M. R. (1978). High pressure phase transformation in $FeF_2$ (rutile). *Geophys. Res. Lett.* **5,** 491.
Welber, B. (1977). Micro-optic system for reflectance measurements at pressures to 70 kbar (application to TTF–TCNQ). *Rev. Sci. Instrum.* **48,** 395.

CHAPTER 8

# MISCELLANEOUS APPLICATIONS

The reality of making ultra-high pressures possible has led researchers to pursue exotic goals that only a few years ago were at best dreams. The advances (reported in Chapter 7) by Mao and Bell (1978) concerning the ultra-high-pressure DAC and by Ruoff and Wanagel (1977) and Ruoff and Chan (1977, 1979) on the diamond indentor DAC have moved some of these dreams into the realm of reality. Attempts at synthesizing metallic hydrogen have been made by other techniques, with no conclusive results. It is interesting to see whether the techniques using the DAC will provide the necessary evidence to substantiate a metallic hydrogen phase. Synthesis of metallic xenon has been claimed by Nelson and Ruoff (1979), who used the diamond indentor DAC.

A metallic superconducting phase of CuCl has been reported. Some of the pitfalls and dangers involved in research of this type will be discussed.

## 1. METALLIC HYDROGEN

Considerable interest in condensed phases under pressure has been manifested since Bridgman's fundamental studies (Bridgman,

1949). The idea of condensing a gas into a solid under pressure and obtaining a metallic phase has gained more attention in recent years. This attention has been spawned by the improved instrumental techniques now available, which make possible such experimentation and the in situ determination of whether a metal has been formed. Although it has long been suspected that any element could become a metal with sufficient pressure, it has only been in the past 10 years that this type of research has become possible. Drickamer and coworkers (Drickamer, 1961; Balchan and Drickamer, 1961; Riggleman and Drickamer, 1962, 1963) lent support to original predictions when they prepared metallic iodine at a pressure of 170 kbar. This was an example of a solid diatomic molecule, $I_2$, being made a metal at high pressure.

The interest shifted to metallic hydrogen and the possibility of making it from gaseous $H_2$ under pressure. Much of the interest in hydrogen corresponded with our increase in knowledge concerning Jupiter, Saturn, and other of the outer planets. Metallic hydrogen is thought to constitute 40% of the mass of the planetary system and to be especially abundant in the planet Jupiter. The notion that the synthesis of metastable, metallic hydrogen was a possibility excited a number of scientists. A better understanding of the properties of metallic hydrogen would improve current models proposed for the outer planets. Additionally, the metal could be useful in a number of other respects, such as in rocket fuel, as an electrical conductor, and possibly even as a superconductor at room temperature. If metallic deuterium could be made, it could be utilized in fusion energy, e.g., for weapons construction and nuclear power production.

Wigner and Huntington (1935) first discussed the possibility of a high-pressure metallic phase of hydrogen. Much effort has been devoted [see McMahan (1978) and references therein] to theoretical calculations for both molecular and metallic hydrogen in an attempt to determine a metal transition pressure. Reviews on this subject have appeared (Ross and Stishkevish, 1977; Ruoff, 1978; McMahan, 1978). Different estimates of a pressure transition to the metallic state have been suggested, but the point remains unclear because no one has successfully made metallic hydrogen to date. The present acceptable consensus suggests that the transition pressure is greater than 1 and less than 3 Mbar.

A number of experimental attempts to make metallic hydrogen have been made, and claims that it had been made have appeared in the literature. Grigorýev et al. (1972) measured the compressibility of hydrogen at pressures from 0.4 to 8 Mbar and observed a sudden decrease in volume at 2.8 Mbar, which they attributed to a metallic phase. The technique used was a dynamic method involving a cylindrical charge of explosive, and the method of calibration of the pressure and the errors involved were not discussed. Hawke et al. (1972) examined hydrogen at a pressure of 2 Mbar with experiments similar to those of the Russian group. However, under compression a rapidly changing magnetic field was used. They reported a density of 1 g cm$^{-3}$. Pressure determination was made indirectly from the amount of compression and was estimated to be 2 Mbar with an error of 1 Mbar. In these experiments involving magnetic compression, no shock wave was produced. Shock-wave techniques, another dynamic method, were also used and pressures of up to 0.9 Mbar reached. In using any of these dynamic methods certain limitations exist. Pressure calibration is uncertain, and it is difficult to make any in situ measurements. Furthermore, it is impossible to recover the metastable metallic hydrogen if it is formed.

Plans to circumvent some of these difficulties have been made. These involve the use of a static method. Mao and Bell (1978) are using their ultra-high-pressure DAC to 1.7 Mbar, and Ruoff (1979) is using an indentor DAC to 3 Mbar. Use of static pressures has been made by Vereschagen et al. (1975), Vereschagen (1976), Kawai et al. (1975), and Spain (1973). Kawai has claimed to have prepared metallic hydrogen (see also Le Neindre et al., 1976). Yakovlev (1979), Vereschagen et al. (1975), and Vereschagen (1976) have claimed to have prepared metallic hydrogen. All claims were based on a resistance drop occurring with pressure. However, no assurances have been made that this was not caused by electrical shorting.

Ruoff (1978) and Nelson and Ruoff (1979) tested their equipment on xenon gas and claimed a conversion of the gas to the metallic state. Small amounts of xenon were condensed on microelectrodes mounted on diamond anvils and pressure was subsequently applied. For a description of this cell see Figs. 5 and 6 in Chapter 2. Under pressure the xenon conducted electricity, and when the pressure was lowered conductivity ceased. They used their nonshorting interdigitated electrode system to make resistance measurements (see Fig. 17, Chapter 2); see also Ruoff and Chan (1979). Mao and Bell (1979), using the ultra-high-pressure DAC, applied pressure to hydrogen gas and obtained solid hydrogen at 25°C and 57 kbar. At 360

# 1. Metallic Hydrogen

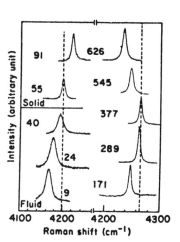

*Fig. 1.* Spectra of the $Q_1(1)$ Raman mode (H–H stretch band) of hydrogen at selected pressures, 9–626 kbar. Instrumentation: excitation by Ar-laser line 488.0 nm, 400 mW; spectral-slit width 5 cm$^{-1}$ for fluid phase and 3.5 cm$^{-1}$ for solid phase. The number by each carve gives the pressure in kilobars. [From Sharma *et al.* (1980).]

kbar, the density rose to 0.6–0.7 g cm$^{-3}$. Sharma (1979) and Sharma *et al.* (1980) have made successful Raman measurements of hydrogen in the pressure range 0.2–630 kbar at room temperature. Figure 1 shows the $Q_1(1)$ Raman band ($\nu_{HH}$) at various pressures. Increases in frequency are noted up to 330 kbar, but decreases occur between 360 and 630 kbar. The results are consistent with a weakening of molecular bonds occurring with pressure. Figure 2 shows the effects of pressure on the rotational bands of solid hydrogen. These bands become diffuse at the solidification point, as expected.

At the present writing, no definite evidence exists that is consistent with a metallic hydrogen phase existing. Certainly, static instrumentation is now available in the diamond indentor DAC and ultrahigh-pressure DAC to reach the proper pressures to compress the gas into a metal. Whether the metal will be metastable is another imponderable. Dr. A. L. Ruoff (1979, p. 10) of Cornell University has written,

> I am often asked whether I believe that metallic hydrogen will be metastable—that is, whether it will persist at ordinary pressures if kept at low temperatures. My answer is and has been: No. My reason is that the metallic state involves long-range forces, and the only examples of metastability involving large energy differences (0.1 eV per atom, say) involve short-range covalent bonds. Moreover, the degree of "unhappiness" of metallic hydrogen—expressed as a free energy of 4 eV per atom at zero pressure—constitutes a tremendous driving force to return to the molecular state. I could be wrong, but I'll be very surprised if I am. I am also asked whether I think hydrogen will be a superconductor at relatively high temperatures (above 100° K and possibly at room temperature), as some theorists have predicted. Again, I have

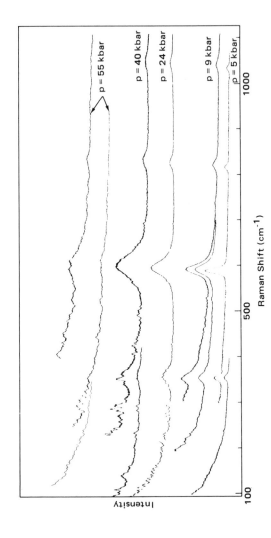

*Fig. 2.* Raman spectra of $n$-$H_2$ in the region of rotational modes at different pressures (488.0-nm $Ar^+$ laser excitation, 200 mW at the sample). Spectral-slit width 10 $cm^{-1}$. The lower traces of the Raman spectra at different pressures were recorded with an amplification of ×0.5. The lowest trace in the low-frequency region in the spectrum of hydrogen at 55 kbar was recorded with a slit width of 5 $cm^{-1}$. [From Ruoff (1979).]

## 2. Metallic Xenon

to say no. Here my reason is that the theorists have calculated the electron behavior of hydrogen metal on the assumption of a static lattice; but in hydrogen the ions (protons) are very light and have appreciable kinetic energy. I expect that accounting for this properly will drastically alter the theoretical predictions.

If I were asked to suggest what will be the highest transition temperature of an element to a metal, I would say that it may be as high as 30°K at pressures of a couple of megabars, and that the metal will be carbon, nitrogen, or oxygen.

Much is still unexplored and unknown in the realm of ultrahigh pressure. There is a long way to go from millions of atmospheres of pressure exerted over dimensions of less than a millionth of a meter to ultrahigh pressures of practical use. But this is partly what makes high-pressure studies an exciting field of research: the possibilities for innovative work are great, and the potential for discovery and application is enormous. Remarkable results are anticipated, hoped for—or perhaps still unimagined.

## 2. METALLIC XENON

Ruoff (1979), Ruoff *et al.* (1977, 1978, 1979), and Ruoff and Nelson (1978) have claimed to have prepared the metallic state of xenon by using their impinging DAC and measuring the resistance drop by the nonshort technique. Special techniques had to be developed to handle gases under pressure in their cell. Figure 3 shows the

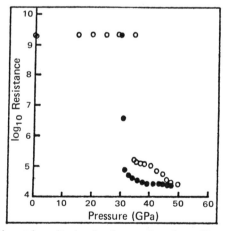

*Fig. 3.* Experimental results showing the creation of metallic xenon. Transitions from insulator to conductor were observed as drops in resistance (as measured on a Keithley 160B multimeter). The temperature was 32 K, and the interdigitated electrode system was used in the measurements. The readings show a drop in resistance by a factor of $10^4$–$10^5$, occurring at about 33 GPa (or 330 kbars). ○, loading; ●, unloading. [From Sharma *et al.* (1980).]

resistance of xenon versus pressure. A sharp drop of $10^4$ to $10^5$ orders in ohms is noted at about 330 kbar. However, Ross and McMahan (1980), on the basis of theoretical computations, have predicted that the xenon should remain an insulator to 1.3 Mbar. Experimental data have refuted the results of Ruoff and co-workers. Syassen (1982) has found no metallization of xenon up to 330 kbar. He observed an onset of a weak ultraviolet absorption at greater than 360 kbar, which was attributed to the indirect $5_p$–$5_d$ interband transition of xenon. At 440 kbar he measured the optical gap as $3.9 \pm 0.3$ eV in agreement with results of Ross and McMahan (1980) and he agreed with the latter in postulating that the metallization of xenon must occur at pressures greater than 1.3 Mbar. See Chan et al. (1982) and the addendum at the end of this chapter.

## 3. CUPROUS CHLORIDE—SUPERCONDUCTOR OR NOT?

The saga of what has come to be called the copper chloride anomaly began during the midseventies. Large diamagnetic susceptibilities suggestive of the Meissner effect at temperatures above 100 K (Rusakov et al., 1975; Chu et al., 1975; Brandt et al., 1978) were observed. A review on the topic appeared in *Physics Today* in 1978. The existence of superconductivity was postulated to explain some of the anomalies (Rusakov et al., 1975; Chu et al., 1978). Other anomalies in compressibility and resistance (Rusakov et al., 1977) and magnetic susceptibility were seen (Chu et al., 1978). Serebryanya (1975) published a complex bulk diagram for CuCl, which differed from phase diagrams for $A^{III} B^V$ and $A^{II} B^{VI}$ compounds (see Phillips, 1973). The usual sequence with increase in pressure is

ZB (zinc blende) semiconductor ⟶ NaCl insulator
⟶ bcc or tetragonal metal

It was suggested that in CuCl the sequence was

ZB semiconductor ⟶ ZB metal phase ⟶ NaCl and metal phase
   40 kbar          60 kbar            ↓ 110 kbar
         metallic phase ⟵
              300 kbar              NaCl phase

The formation of the zinc blende metal phase at 40 kbar is not accompanied by any detectable structural change, which is suspi-

## 3. Cuprous Chloride—Superconductor or Not?

**Table I**

*Crystallographic and Volume Data on CuCl*

| Pressure (kbar) | Phase | Cell constants (Å) | Volume (Å³) |
|---|---|---|---|
| 0 | ZB[a] | $a = 5.416$ | 39.172 |
| 44 ± 2 | ZB | $a = 5.272 \pm 0.006$ | 36.638 |
| 44 ± 2 | Tetragonal | $a = 5.270 \pm 0.01$<br>$c = 4.688 \pm 0.002$ | 32.558 |
| 82 ± 2 | Tetragonal | $a = 5.191 \pm 0.015$<br>$c = 4.619 \pm 0.03$ | 31.056 |
| 82 ± 2 | NaCl | $a = 4.96$ | 30.506 |
| 123 ± 2 | NaCl | $a = 4.90$ | 29.412 |

[a] ZB represents zinc blende.

cious. As a consequence, the sequence appeared to be implausible, according to Blount and Phillips (1978). In 1979, Piermarini et al. (1979) examined CuCl at pressures of up to 1.2 Mbar by using a DAC. Electrical measurements were made to 0.55 Mbar, and x-ray studies were conducted as well. In the pressure domain measured, no anomalous dependency or electrical resistance was found. No dramatic changes in optical transmission were observed up to 100 kbar. Four phases were identified. At 84 kbar both tetragonal and NaCl-type phases are present. At 96 kbar only the NaCl phase exists. No evidence for a metal transition was found up to 1.2 Mbar. Table I summarizes the results of Piermarini et al. (1979).

Blount and Phillips (1978) explained the dilemma in terms of a disproportionation reaction occurring at high pressures, such as

$$2 \text{ CuCl} \longrightarrow \text{Cu}^\circ + \text{CuCl}_2 \qquad (1)$$

Precedents for such reactions were found by Bell et al. (1976) (see Chapter 7). At higher pressures the copper aggregates can be absorbed into the rock salt phase, which gives an opaque material. Piermarini et al. (1979) consider the following reactions as possibly occurring:

$$2 \text{ CuCl} \longrightarrow \text{CuCl} + \text{Cu}^\circ \qquad (2)$$

$$3 \text{ CuCl} \longrightarrow \text{Cu}^+ \text{Cu}^{++} + \text{Cl}_3^- + \text{Cu}^\circ \qquad (3)$$

$$4 \text{ CuCl} \longrightarrow \text{Cu}_2^+ + \text{Cl}_4^- + \text{Cu}^\circ \qquad (4)$$

Each of these reactions produces small amounts of copper metal, which could account for the cutoff of visible light transmission observed. Wilson (1978) proposed that ordered segregation of copper metal in CuCl was possible.

Confirmation for the preceding conclusions has come from Vezzoli and Bera (1981), Vezzoli (1982), and Skelton *et al.* (1980). No evidence of a superconductor was found, and evidence for formation of metallic copper was provided. Hinze (1978) found optical and x-ray evidence for formation of metallic copper under pressure for $Cu_2S$ and $Cu_2Se$.

Present consensus is in agreement that the anomalous bulk metallic phase of CuCl near 40 kbar observed by Serebryanaya *et al.* (1975) is nothing more than small amounts of copper metal from the disproportionation reactions (1)–(4). This would explain the electrical and structural data, but the Meissner effect remains inexplicable and appears to be doubtful.

The copper chloride dilemma illustrates the dangers involved in subjecting materials to high pressures. It becomes possible to form the free metal of the compound under surveillance by disproportionation reactions under pressure. Ruoff (1979) has cited the dangers in measuring resistance under pressure when shorting occurs (see section on metallic hydrogen). It is obvious that wrong conclusions can be reached.

## ADDENDUM

Chan *et al.* (1982) have found that the pressure in solid Xe when it exhibits electrical conductivity is much greater than is indicated by the results of Nelson and Ruoff (1979). New data show pressure to be over 1 Mbar (previously reported as 330 kbar) and agree with the 1.3-Mbar value predicted by Ross and McMahan (1980).

## REFERENCES

Balchan, A. S., and Drickamer, H. G. (1961). *J. Chem. Phys.* **34,** 1948.
Bell, P. M., Mao, H. K., Weeks, R. A., and Van Valkenburg, A. (1976). *Annu. Rep. Geophys. Lab., Washington, D.C.* **75,** 515.
Blount, E. I., and Phillips, J. C. (1978). *J. Less Common Met.* **62,** 457.
Brandt, N. B., Kurshinnkov, B. V., Rusakov, A. P., and Gomionov, M. V. (1978). *Pisma Zh. Eksp. Teor. Fiz.* **27,** 37.

## References

Bridgman, P. W. (1949). "Physics of High Pressure." Bell and Son, London.
Chan, K. S., Huang, T. L., Grzybowski, T. A., Whetten, T. J., and Ruoff, A. L. (1982). *Phys. Rev. B* **26,** 7116.
Chu, C. W., Early, S., Geballe, T. H., Rusakov, A. P., and Schwall, R. E. (1975). *J. Phys. C* **8,** L241.
Chu, C. W., Rusakov, A. P., Huang, S., Early, S., Geballe, T. H., and Huang, C. Y. (1978). *Phys. Rev. B* **18,** 2116.
Drickamer, H. S. (1961). "Progress of Very High Pressure Research." Wiley, New York.
Grigorýev, F. V., Kormer, S. B., Mikhaylova, O. L., Tolochko, A. P., and Urlin, V. D. (1972). *Zh. Eksp. Teor. Fiz. Pisma Red.* **5,** 286.
Hawke, R. S., Duerre, D. E., Huebel, J. G., Keeler, R. N., and Klapper, H. (1972). *Phys. Earth Planet. Inter.* **6,** 44.
Hinze, E. (1978). *Acta Crystallog. Part A* **34,** S343.
Kawai, N., Togaya, M., and Mishima, O. (1975). *Proc. Jpn. Acad.* **51,** 630.
LeNeindre, B., Suito, K., and Kawai, N. (1976). *High Temp.–High Pressures* **8,** 1.
McMahan, A. K. (1978). *In* "High Pressure Low Temperature Physics" (C. W. Chu and J. A. Wollam, eds.), pp. 21–42. Plenum, New York.
Mao, H. K., and Bell, P. M. (1978). *Science* **200,** 1145.
Mao, H. K., and Bell, P. M. (1979). *Science* **203,** 1004.
Nelson, D. A., and Ruoff, A. L. (1979). *Phys. Rev. Lett.* **42,** 383.
Phillips, J. C. (1973). "Bonds and Bands in Semiconductors." Academic Press, New York.
*Phys. Today* (September 1978), 17–19.
Piermarini, G. J., Mauer, F. A., Block, S., Jayaraman, A., Geballe, T. H., and Hull, G. W. (1979). *Solid State Commun.* **32,** 272.
Riggleman, B. M., and Drickamer, H. G. (1962). *J. Chem. Phys.* **37,** 446.
Riggleman, B. M., and Drickamer, H. G. (1963). *J. Chem. Phys.* **38,** 2721.
Ross, M., and McMahan, A. K. (1980). *Phys. Rev. B* **21,** 1658.
Ross, N., and Stishkevish, C. (1977). Molecular and Metallic Hydrogen, Report R-2056-ARPA, Rand Corp., Santa Monica, California.
Ruoff, A. L. (1978). *In* "High Pressure Low Temperature Physics" (C. W. Chu and J. A. Wollam, eds.), pp. 1–20. Plenum Press, New York.
Ruoff, A. L. (1979). *Cornell Q. Eng.* **14,** 2–10.
Ruoff, A. L., and Chan, K. S. (1979). *High Pressure Sci. Technol. AIRAPT Conf., 6th, 1977,* **1,** 779–784. [Also Cornell Univ. Materials Science Center Report No. 2854 (1977). Cornell Univ., Ithaca, New York.]
Ruoff, A. L., and Nelson, D. A. (1978). *Chem. Eng. News* **46**(47), 22.
Ruoff, A. L., and Wanagel, J. (1977). Cornell Univ. Materials Science Center Report No. 2892. Cornell Univ., Ithaca, New York.
Rusakov, A. P., Laukhin, V. W., and Lisovskü, Y. A. (1975). *Phys. Status Solids B* **71,** K191.
Rusakov, A. P., Grigoryan, S. G., Omelchenko, A. O., and Kalyshevich, A. E. (1977). *Sov. Phys., JETP, Engl. Trans.* **45,** 380.
Serebryanya, N. R., Popova, S. V., and Rusakov, A. P. (1975). *Fiz. Tverd Tela, 1975* **17,** 2772.
Sharma, S. K. (1979). *Raman Newsletter* (12415), 2.
Sharma, S. K., Mao, H. K., and Bell, P. M. (1980). *Phys. Rev. Lett.* **44,** 886.

Skelton, E. F., Webb, A. W., Rachford, F. J., and Taylor, P. C. (1980). *Phys. Rev. B* **21,** 5289.
Spain, I. (1973). *Science* **180,** 399.
Syassen, K. (1982). *Phys. Rev. B* **25,** 6548.
Vezzoli, G. C. (1982). *Phys. Rev. B* **26,** 4140.
Vezzoli, G. C., and Bera, J. (1981). *Phys. Rev. B* **23,** 3022.
Vereschagen, L. F. (1976). *New Sci.* **71**(1016), 477.
Vereschagen, L. F., Yakovlev, E. N., and Timofeev, Yu, A. (1975). *JETP Lett.* **21,** 85.
Wigner, E., and Huntington, H. B. (1935). *J. Chem. Phys.* **3,** 764.
Wilson, J. A. (1978). *Philos. Mag. B* **38,** 427.
Yakovlev, Y. (1976). *New Sci.* **71**(1016), 478.

# ADDITIONAL BIBLIOGRAPHY

This bibliography relates to material that has appeared since the book was submitted to the publisher. The material is categorized according to the topics discussed in the chapters in the book.

The following papers were presented at the Ninth AIRAPT International High-Pressure Conference, State University of New York, Albany, July 24–29, 1983 (abstracts).

*Chapter 2. Instrumentation*

*Synchrotron (X-Ray)*

Grzybowski, T. A., and Ruoff, A. L. X-ray and optical studies of BaSe and BaTe.
Huang, T., and Ruoff, A. L. High pressure phases of HgS, HgSe and HgTe.
Huang, T., and Ruoff, A. L. High pressure x-ray diffraction study on cesium iodide.
Iwasaki, H., and Kikegawa, T. Simple cubic structure as a stable form of phosphorus under high pressure.
Keeler, R. N., and Rhodes, C. K. Laser generated x-ray and extreme UV-photon sources for static and dynamic high pressure studies.
Kraft, A., Vollstädt, H., Kühn, G., and Müller, W., Behavior of $CuGaSe_2$ and $CuGaTe_2$ under high pressure.

Manghnani, M. H., Ming, L. C., Balogh, J., Jamieson, J. C., Qadri, S. B., Webb, A. W., and Skelton, E. F. Simultaneous P-V-T measurements up to 20 GPa and 900 K using synchrotron radiation.
Qadri, S. B., Skelton, E. F., Webb, A. W., Wolf, S. A., Elam, W. T., and Rek, Z. Investigation of the structure and compressibility of a number of chalcopyrite compounds.
Schilling, J. C. Magnetism at high pressure.
Shimomura, O., Yamaoka, S., Yagi, T., Wakatsubi, M., Tsuji, K., Fukunaga, O., Kawamura, H., Aoki, K., and Akimoto, S. Multi anvil apparatus for synchrotron radiation.
Skelton, E. F., Webb, A. W., Elam, W. T., Wolf, S. A., Qadri, S. B., Huang, C. Y., Chaikin, P. M., Lacoe, R. C., and Gschniedner, K. A. Structural studies at elevated pressure and reduced temperatures using synchrotron radiation.
Webb, A. W., Qadri, S. B., Skelton, E. F., Elam, W. T., and Rek, Z. Linear compressibility studies of selected transition metal dichalogenides.
Will, G., Lawterjung, J., Schmidt, H., and Hinze, E. The bulk moduli of 3d-transition element pyrites measured with synchrotron radiation in a new belt type apparatus.

## Conductivity

Drickamer, H. G. High pressure studies of semiconductor–electrolyte interfaces.
Dunn, K. J. Electrical properties of solids at high pressure.
Schloessin, H. H., and Govindarajan, R. Electrical conductivity of magnetite at temperatures from 90–900 K at pressures from 0 to 5.6 GPa.
Sundquist, B., and Lundberg, B. Contactless methods for measuring electrical resistivity under pressure.
Vezzoli, G. E. Electrical and optical behavior at high pressure and temperature of silver nitrate, cobalt nitrate hydrate and sodium nitrate.

## EPR

Barnett, J. D., Decker, D. L., and Ellwanger, M. H. EPR study of $BaTiO_3$-$Fe^{3+}$ in the tetragonal ferroelectric phase at high pressure.
MacCrone, R. K., and Homan, C. G. ESR measurements in pressure quenched CdS containing Cl.

## Brillouin Scattering

Grimsditch, M., and Rahman, A. Elastic moduli of $H_2O$: Liquid ice VI and ice VII.

## Mössbauer Spectroscopy

Abd-Elmeguid, M. M., and Micklitz, H. Application of high pressure Mössbauer spectroscopy to study the local environment of amorphous metals.

Suito, K., Tsutsui, Y., Nasu, S., Onodera, A., and Fujita, F. E. Mössbauer effect study of γ-form of $Fe_2SiO_4$.

## Chapter 4. Inorganic Compounds

Barnett J. D., Decker, D. L., and Ellwanger, M. H. EPR study of $BaTiO_3$–$Fe^{3+}$ in the tetragonal ferroelectric phase at high pressures.
Block, S., and Piermarini, G. Chemistry of CdS materials containing Cl which exhibit magnetic and electric anomalies.
Boehler, R. Thermodynamic properties of the alkali metals at high pressures and high temperatures.
Chandrasekhar, M., and Collins, T. C. Spin dependence calculations of possible excitonic superconductivity states in CuCl and CdS.
Chattopadhyay, T., Werner, A., and von Schnering, H. G. Pressure induced phase transition in IV–VI compounds.
Collins, T. C. Theoretical models for collective behavior in CuCl and CdS.
Cote, P., Capsimalis, G., and Moffatt, W. C., Thermal and x-ray analyses of pressurized CdS containing chlorine impurities.
Decker, D. L., and Zhao, Y. X. Change in the $BaTiO_3$ phase transition to 40 kbar.
Etters, R. D., Chandrasekharan, V., and Kobashi, K. Prediction of high pressure phase transitions in solid $N_2$ and the pressure dependence of the intramolecular vibrational frequencies.
Ewald, A. H. High temperature high pressure absorption spectra of uranyl solution under shock conditions.
Hanson R. C., and Katz, A. Solid hydrogen chloride at high pressures.
Homan, C. G., Laojindapun, K., and MacCrone, R. K. Lambda anomalies in dielectric properties in pressure quenched CdS containing Cl.
Homan, C. G., and MacCrone, R. K. Evidence for collective behavior in pressure quenched CdS containing Cl.
Jayaraman, A., Remeika, J. P., and Katiyan, R. S. High pressure Raman scattering studies on $BaTiO_3$ and $PbTiO_3$.
Klug, D. D., and Whalley, E. The pressure dependence of the infrared spectrum of HDO in $D_2O$ to 139 kilobar.
Kobashi, K., and Etters, R. D. Lattice dynamics of solid $I_2$ under high pressure.
MacCrone, R. K., and Homan, C. G. ESR measurements in pressure quenched CdS containing Cl.
Menoni, C. S., Jing-Zhu Hu, and Spain, I. L. Germanium and silicon under high pressure.
Ming, L. C., and Manghnani, M. H. In situ high pressure transformation studies on the monoclinic $ZrO_2$ and $CrF_2$.
Nam, S. B., Chang, Y., and Reynolds, D. C. Evidence for superconductivity in pressure quenched CdS.
Neilson, G. W. The effect of pressure on the structure of light and heavy water.
Piermarini, G. J., Munro, R. G., and Block, S. Metastability in $H_2O$ and $D_2O$ at high pressure.
Ribarsky, M. W. Local field effects on exciton coupling in CuCl and CdS under pressure.

Scholz, W., Homan, C. G., Moffatt W. C., and MacCrone, R. K. Superconductivity in pressure changed palladium hydride.

Schouten, J. A., van den Bergh, L. C., and Trappeniers, N. S. Investigation of the phase equilibria in the system neon–xenon using a diamond-anvil system.

Strössner, K., Henkel, W., Werner, A., and Hochheimer, H. D. High pressure study of the alkali cyanides.

Vögele, H. P., and Tödheide, K. Photoluminescence of of *bis*-tetrabutyl–ammonium tetrabromomanganate to 440 K and 3 kbar.

Xu, J., Mao H. K., and Bell, P. M. Infrared measurements of solid $H_2$ at pressures greater than 0.5 Mbar.

## Chapter 5. Coordination Compounds

Stankowski, J., and Krupski, M. Phase diagrams of hexammines studied by EPR under high pressure.

van Eldik, R., Brect, E. L. J., Doss, R., Mohr, R., and Kelm, H. Mechanistic information from the effect of pressure on fast substitution reactions of inorganic complexes in aqueous solutions.

Vezzoli, G. C. Electrical and optical behavior at high pressure and temperature of silver nitrate, cobalt nitrate hydrate, and sodium nitrate.

Weber, W., Küster, U., van Eldik, R., and Kelm, H. Effect of pressure on the photosolvolysis reactions of a series of pentaamminerhodium(III) complexes in solution.

## Chapter 6. Organic and Biological Compounds

Fabre, D., Thiery, M. M., and Kobashi, K. Spectroscopic (Raman) evidence of a new high pressure phase in solid $CD_4$.

Green, E., Lee, L., Mitchell, A., Ree, F., Tipton, R., and van Thiel, M. Detonation product EOS patterns for several explosives.

Moore, D. S., Schmidt, S. C., Shiferl, D., and Shaner, J. W. Single-pulse coherent Raman spectroscopy in shock–compressed benzene.

Scaife, W. G. Dielectric behavior of freezing *n*-decane and *n*-pentadecane.

Trappeniers, N. J. Phase transitions in solid methane up to 60 kbar.

Wong, P. T. T., and Mantsch, H. H. Pressure and temperature effects on the Raman spectrum or aqueous micellar solution of sodium oleate.

Wood, S. D., and Bean, V. E. Compression of $CCl_4$ at high pressures.

## Chapter 7. Special Applications

### Geochemical and Geophysical Applications

Endo, S., Akahama, Y., Wakatsuki, M., Nakamura, T., and Tomii, Y. Synthesis of the single crystal of stishovite.

Ito, E. The system $MgO–FeO–SiO_2$ at ultra-high pressures.

# Additional Bibliography

Lebedev, T. S. Some problems of experimental P–T studies of physical properties of lithosphere's mineral matter.
Le Neindre, B., and Petitet, J. P. Thermodynamic properties of molten silicates under pressure: Geophysical applications.
Manghnani, M. H., and Matsui, T. Effect of temperature on the pressure derivatives of single-crystal elastic constants of pure forsterite ($Mg_2SiO_4$).
Togaya, M. High pressure generation and phase transformations of $SiO_2$ and $GeO_2$.
Weng, K., Xu, J., Mao, H. K., and Bell, P. M. High pressure infrared spectra of mantle materials (olivine, pyroxene).

## Conductive Field

Schirber, J. E. Low temperature pressure studies of organic superconductors.
Weitz, D. A., King, H. E., Stokes, J. P., Chung, J. C., and Bloch, A. N. High-pressure resonance Raman scattering from the l-D organic conductor HMTSF–TCNQ.
Wolf, S. A., Huang, C. Y., Chaikin, P. M., Fuller, W. W., Lacoe, R. C., Luo, H. L., and Wudl, F. The role of pressure in understanding the anomalous superconductivity in $EuMo_6S_8$ and $(TMTSF)_2FSO_3$.

## Lubricant Studies

Gardiner, D. J., Baird, E. M., Gorvin, A. C., and Dare-Edwards, M. P., Raman spectra of fluids in elastohydrodynamic contacts.

The following papers are references from the literature pertaining to the topics discussed in this book.

## Chapter 2. Instrumentation

### DAC

Adams, D. M., and Shaw, A. C. (1982). A computer aided design study of the behavior of diamond anvils under stress, *J. Phys. D* **15**, 1609.
Bell, P. M. (1983). Static ultrahigh-pressure research, *Rev. Geophys. Space Phys.* **21**, 1394–1399, and references therein.
Chan, K. S., Huang, T. L., Grzybowski, T. A., Whetten, T. J., and Ruoff, A. L. Pressure concentrations due to plastic deformation of thin films or gaskets between anvils, *J. Appl. Phys.* **53**, 6607.
Jian, X. and Jingzhu, H. (1982). Technique of the diamond anvil cell, *Wuli* **11**, 82.
Mao, H. K., Bell, P. M. Xu, J., and Wong, P. T. T. (1982–1983). High-pressure Fourier transform spectroscopy, *Annu. Rep. Geophys. Lab. Washington, D.C.* **82**, 419–421.

Mao, H. K., Hadidiacos, C. G., Bell, P. M., and Goettel, K. A. (1982–1983). Automated systems for heating and for spectral measurements in the DAC, *Annu. Rep. Geophys. Lab., Washington, D.C.* **82**, 421–424.

Mohammed, K. (1982). Uniaxial stress splitting of photoluminescence transitions in yellow luminescing brown diamonds, *J. Phys. C* **15**, 1789.

Munro, R. G., Block, S., Mauer, F. A., and Piermarini, G. J. (1982). Radial distribution studies in a diamond anvil pressure cell, *J. Appl. Phys.* **53**, 7080.

Takano, K., Wakatsuki, M., and Tabata, K. (1982). Application of rapid spectroscopic measurement to diamond anvil high pressure cell III, *23rd High Pressure Conf., Japan, November 18–20, 1982*, Abstr. 14.

## Optical Cell

Holland, L. R., Harris, R. P., and Smith, R. (1983). High temperature, high pressure optical cells, *Rev. Sci. Instrum.* **54**, 993.

Rogers, V. E., and Angell, C. A. (1983) An inexpensive high pressure optical absorption cell for IR–VIS–UV studies, *J. Chem. Educ.* **60**, 602.

## Optical Windows

Perry, S., Sharko, P. T., and Jonas, J. (1983). Technique for measuring the amount of pressure induced polarization scrambling by optical windows in high pressure light scattering cells, *Appl. Spectrosc.* **37**, 340.

## X-Ray Spectroscopy

Baublitz, M. J., and Ruoff, A. L. (1983). X-ray diffraction data from the high pressure phase of aluminum antimonide, *J. Appl. Phys.* **54**, 2109.

Finger, L. W., Hazen, R. M., Xu, G., Mao, H. K., and Bell, P. M. (1981). Structure and compression of crystalline argon and neon at high pressure and room temperature, *Appl. Phys. Lett.* **39**, 892.

Fukizawa, A., and Akimoto, S. (1982). Pressure measurements in the high pressure and temperature x-ray experiment using NaCl and gold, *23rd High Pressure Conf., Japan, November 18–20, 1982*, Abstr., p. 72.

Kikegawa, T., and Iwasaki, H. (1983). X-ray diamond anvil press for use at high temperatures, *Rev. Sci. Instrum.* **54**, 1023.

Mao, H. K., Bell, P. M., and Weng, K. (1982–1983). Position-sensitive x-ray diffraction system for high-pressure experiments, *Annu. Res. Geophys. Lab., Washington, D.C.*, **82**, 424–428.

Roehler, J., Kappler, J. P., and Krill, G. (1983). High pressure device for use in low energy x-ray absorption spectroscopy, *Nucl. Instrum. Meth. Phys. Res.* **208**, 647.

## Mössbauer Spectroscopy

Nasu, S., Nagatomo, S., Kawamura, T., Endo, S., and Fujita, E. (1982). Mossbauer spectroscopy under high pressure, *23rd High Pressure Conf., Japan, November 18–20, 1982,* Abstr., p. 36.

Tsutsui, Y., Suito, K., Nasu, S., Onodera, A., and Fujita, E. (1982). Fe Mossbauer effect on high pressure phase $\gamma$-$Fe_2SiO_4$, *23rd High Pressure Conf., Japan, November 18–20, 1982,* Abstr., p. 34.

## ERP

Petrakovskaya, E. A., Velichko, V.V., Krygin, I. M., Petrov, S. B., and Falaleeva, L. G. (1983). EPR study of manganese(II) impurity centers in cesium cadmium chloride ($CsCdCl_3$) at hydrostatic pressure, *Fiz. Tverd. Tela (Leningrad)* **25,** 862.

## NMR

Schuele, P. J., and Schmidt, V. H. (1982). High pressure, low temperature apparatus for NMR study of phase transitions, *Rev. Sci. Instrum.* **53,** 1724.

Shimokawa, S., and Yamada, E. (1983). A carbon 13 NMR probe for high temperature and high pressure experiments, *J. Magn. Resonance* **51,** 103.

Shimokawa, S., Yamada, E., and Makino, K. (1983). A simple pressurized high temperature proton NMR apparatus for thermal degradation of polymers, *Bull. Chem. Soc. Jpn.* **56,** 412.

Tubino, M., and Merbach A. E. (1983). High pressure NMR kinetics. 17. Variable temperature and variable pressure proton NMR studies of dimethyl sulfide exchange on trans *bis*(dimethyl sulfide)dichloro palladium(II) on various solvents, *Inorg. Chem* **71,** 149.

## Brillouin Scattering

Benoit, J. P., Thomas, F., and Berger, J. (1983). Effect of hydrostatic pressure on Brillouin scattering spectra of thiourea, *J. Phys. (Les Ulis. Fr).* **44,** 841.

Shimidzu, H., Kumajawa, T., Bassett, W., and Brody, E. M. (1982). Brillouin scattering and lattice dynamics of fosterite under high pressure, *23rd High Pressure Conf., Japan, November 18–20, 1982,* Abstr., p. 8.

Shimidzu, H., Kumazawa, T., Brody, E. M., Mao, H. K., and Bell, P. M. (1983). Acoustic velocity measurements for various directions on solid $n$-deuterium ($n$-$D_2$) at 136 kbar and 300 K by Brillouin scattering in a diamond anvil cell, *Jpn. J. Appl. Phys.* **22,** 52.

Sooryakumar, R., and Simmonds, P. E. (1983). High resolution resonant Brillouin scattering and the effect of stress on gallium arsenide, *Phys. Rev. B* **27,** 4978.

## Neutron Scattering

Axe, J. D., McWhan, D. B., and Nigrey, P. J. (1983). Neutron scattering study of hydrostatic pressure effects in mercury hexafluoroarsenate ($Hg_{3-\delta}AsF_6$), *Phys. Rev. B*, **27**, 3845.

## Synchrotron (X-Ray)

Buras, B. (1983). High pressure research with synchrotron radiation, *Nucl. Instrum. Meth. Phys. Rev.* **208**, 563.

Qadri, S. B., Skelton, E. F., and Webb, A. W. (1983). High pressure studies of germanium using synchrotron radiation, *J. Appl. Phys.* **54**, 3609.

## Chapter 3. Pressure Calibration

Ming, L. C., Manghnani, M. H., Balogh, J., Qadri, S. B., Skelton, E. F., and Jamieson, J. C. (1983). Gold as a reliable internal pressure calibrant at high temperatures, *J. Appl. Phys.* **54**, 4390.

Minguang, Z., Xu, J., and Huiru, B. (1982). d-Orbital theory for ruby and its application to high pressure spectra. *Scientia Sinica A* **25**, 1066.

## Chapter 4. Inorganic Compounds

### General

Adams, D. M., and Hatton, P. D. (1983). Vibrational spectroscopy at high pressures, Part 42, The Raman spectrum of mercury(II) Thiocyanate, *J. Chem. Soc. Perkins Trans. 2* **79**, 695.

Adams, D. M., and Hatton, P. D. (1983). Vibrational spectroscopy at high pressures, Part 43, *J. Phys. Chem.* **16**, 3349.

Adams, D. M., and Hatton, P. D. (1983). Spectroscopy at very high pressures. 45. Structural phase transitions in $A_2HgI_4$ (A–Ag, Cu, Tl, $\frac{1}{2}$ Pb) investigated by Raman scattering, *J. Raman Spectrosc.* **14**, 154.

Adams, D. M., Hatton, P. D., and Taylor, I. D. (1983). Vibrational spectroscopy at high pressures. 44. $Tl_2CO_3$, *J. Raman Spectrosc.* **14**, 144.

Asaumi, K., Suzuki, T., and Mori, T. (1983). High-pressure optical absorption and x-ray diffraction studies in RbI and KI approaching the metallization transition, *Phys. Rev. B* **28**, 3529–3533.

Batlogg, B., Jayaramah, A., Vancleve, J. E., and Maines, R. G. (1983). Optical absorption, resistivity, and phase transformation in cadmium sulfide at high pressure, *Phys. Rev. B* **27**, 3920.

Beck, H. P., Limmer, A., Denner, W., and Schulz, H. (1983). Influence of high hydrostatic pressure on the crystal structure of barium flouride chloride in the pressure range up to 6.5 GPa, *Crystallogr. Part B* **39,** 401.

Chikamatsu, M., Yamazaki, H., Yamagata, K., and Abe, H. (1983). Pressure effects on magnetic phase transitions in $(C_2H_5NH_3)_2CuCl_4$, *J. Magn. Materials* **31,** 1191.

Gesi, K. (1982). Pressure induced ferroelectricity in $[N(CD_3)_4]_2ZnCl_4$, *J. Phys. Soc. Jpn.* **51,** 1043.

Gesi, K., and Ozawa, K. (1982). Effect of hydrostatic pressure on the phase transitions in ferroelectric $[N(CH_3)_4]_2CuBr_4$, *J. Phys. Soc. Jpn.* **51,** 2205.

Gorczyka, I. (1982). Effect of hydrostatic pressure on the band structure of indium antimonide. A pseudopotential calculation, *Phys. Status Solidi B* **112,** 97.

Heyns, A. M., and Clark, J. B. (1983). Effects of pressure on Raman spectra of solids. 1. $KClO_3$, *J. Raman Spectrosc.* **14,** 342.

Homan, C. G., Kendall, D., and MacCrone, R. K. (1981). Electrical and magnetic properties of pressure quenched CdS. *In* "Physics of Solids under High Pressures," 407. North-Holland Publ., Amsterdam, Netherlands.

Huang, T., and Ruoff, A. L. (1983). Pressure induced phase transition of mercury(II) sulfide, *J. Appl. Phys.* **54,** 5459.

Jing-zhu, H. (1981). Observation of high pressure pnase transition in iodine and sulfur with the use of a diamond anvil device, *Chin. Phys.* **1,** 594.

Kawamura, T., Endo, S., Kobayashi, M., Cho, K., and Narita, S. (1982). Optical measurements of $\alpha$-$Fe_2O_3$ and $AgBr_{1-x}Cl_x$ under high pressure, *23rd High Pressure Conf., Japan, Nov. 18–20, 1982,* Abstr., p. 12.

Knittle, E., and Jeanloz, R. (1984). Structural and bonding changes in CsI at high pressures, *Science* **223,** 53.

Kobashi, K., and Etters, R. D. (1983). Lattice dynamics of solid diatomic iodine under high pressures, *J. Chem. Phys.* **79,** 3018.

Kultz, W., and Rehaber, E. (1983). Light scattering studies of RbCN under hydrostatic pressure, *Phys. Rev. B* **28,** 2114.

Kuroda, N., Iwabuchi, T., and Nishina, Y. (1983). Raman scattering and Fundamental absorption in red $HgI_2$ under hydrostatic pressure, *J. Phys. Soc. Jpn.* **52,** 2419.

Mammone, J. F., and Sharma, S. K. (1978–1979). Raman Study of $GeO_2$ in crystalline and glassy states at high pressures, *Annu. Geophys. Lab., Washington, D.C.,* **78,** 640.

Okai, B. (1983). Mode instability in sodium chloride structure under pressure, *J. Phys. Chem. Jpn.* **52,** 2289.

Shanker, J., and Singh, K. (1982). Analysis of crystal binding and Grüneisen parameter in cesium halides, silver halides and alkaline earth oxides, *Phys. Status Solidi B* **113,** 737.

Smolander, K. J., (1983). On the high-pressure polymorphism of cuprous chloride, *J. Phys. Chem.* **16,** 3673.

Wada, M., Shici, H., Sawada, A., and Ishibishi, T. (1982). The uniaxial stress effect on normal incommensurate phase transition in potassium selenate ($K_2SeO_4$). Observations of the amplitude mode by Raman scattering, *J. Phys. Soc. Jpn.* **51,** 3245.

Wada, M., Takeda, K., Ohtani, A., Shimomura, O., Onodera, A., and Haseda, T. (1982). Effect of pressure on phase transition of $CoCl_2 \cdot 6H_2O$, *23rd High Pressure Conf., Japan, November 18–20, 1982,* Abstr. P. 40.

Werner, A., Hochheimer, H. D., Stoessner, K., and Jayaraman, A. (1983). High-pressure x-ray diffraction studies on $HgTe_2$ and $HgS_2$ to 20 GPa, *Phys. Rev. B* **28**, 330.
Yagi, T., Suzuki, T., and Akimoto, S. (1983). New high pressure polymorphs in sodium halides, *J. Phys. Chem. Solids* **44**, 135.
Yoder, S., Collela, R., and Weinstein, A. (1982). Valence electron density in silicon and indium antimonide under high pressure by x-ray diffraction, *Phys. Rev. Lett.* **49**, 1438.

## Ices

Chandrasekharan, V., Etters, R. D., and Kobashi, K. (1983). Calculation of high pressure phase transitions in solid nitrogen and the pressure dependence of intramolecular mode frequencies, *Phys. Rev. B* **28**, 1095.
Cromer, D. T., Schiferl, D., LeSar, R., and Mills, R. L. (1983). Room temperature structure of CO at 2.7 and 3.6 GPa, *Acta Crystallogr. Part C* **39**, 1144.
Etters, R. D., and Helmy, A. (1983). Pressure dependence of intramolecular mode frequencies in solid nitrogen, oxygen and carbon dioxide, *Phys. Rev. B* **27**, 6539.
Etters, R. D., Schilling, J. S., and Bach, H. (1981). Intramolecular vibrational mode shifts with pressure in solid $CO_2$, $N_2$ and $O_2$. In "Physics of Solids under High Pressures," p. 385. North-Holland Publ., Amsterdam, Netherlands.
Johannsen, P. G., and Holzapfel, W. B., (1983). Effect of pressure on Raman spectra of solid bromine, *J. Phys. Chem.* **16**, 1961.
Levesque, D., Weis, J. J., and Klein, M. L. (1983). New high pressure phase of solid helium-4 in bcc, *Phys. Rev. Lett.* **51**, 670.
Mao, H. K., Xu, J., and Bell, P. M. (1982–1983). Pressure-induced infrared spectra of $H_2$ to 542 kbar, *Annu. Rep. Geophys. Lab., Washington, D.C.* **82**, 366–372.
Nicol, M., and Syassen, K. (1983). High pressure optical spectra on condensed oxygen, *Phys. Rev. B* **28**, 1201.
Nose, S., and Klein, M. L. (1983). Structural transformation in solid nitrogen at high pressure, *Phys. Rev. Lett.* **50**, 1203.
Prikhot'ko, A. F., Pikus, Y. G., Salivon, A. I., and Shanskii, L. I. (1983). Effect of pressure on the absorption spectra of solid oxygen, *Ukr. Fiz. Zh.* **28**, 615.
Ross, M., Ree, F. H., and Young, D. A. (1983). The equation of state of molecular hydrogen at very high density, *J. Chem. Phys.* **79**, 1487.
Swanson, B. I., Babcock, L. M., Schiferl, D., Moody, D. C., Mills, R. L., and Ryan, R. R. (1982). Raman study of $SO_2$ at high pressure: Aggregation, phase transformation and photochemistry, *Chem. Phys. Lett.* **91**, 393.
Xu, J., Mao, H. K., Finger, L. W., Bell, P. M., and Hazen, R. M. (1981). Interatomic potential for solid argon and neon at high pressures. In "Physics of Solids under High Pressures," p. 139. North-Holland Publ., Amersterdam, Netherlands.

## Chapter 5. Coordination Compounds

Adams, D. M., and Ekejiuba, I. O. C. (1982). A second order phase transition in $Mn(CO)_5Br$, *J. Chem. Phys.* **77**, 4793.

Adams, D. M., and Ekejiuba, I. O. (1983). Vibrational spectroscopy at high pressures, XL. A Raman study of manganese rhenium decacarbonyl, *J. Chem. Phys.* **78**, 5408.
Gesi, K. (1983). Ferroelectric region of $(N(CH_3)_4)_2FeCl_4$ in the pressure temperature phase space, *J. Phys. Soc. Jpn.* **52**, 3322.
Meissner, E., Koeppen, H., Spiering, H., and Guetlich, P. (1983). The effect of low pressure on a high spin low spin transition, *Chem. Phys. Lett.* **95**, 163.
Ogata, F., Kambara, T., Gondaira, K. I., and Sasaki, N. (1983). Pressure-induced high spin low spin transition in transition metal compounds, *J. Magn. Materials* **31**, 123.

## Chapter 6. Organic and Biological Compounds

Backer, M., Haefner, W., and Kiefer, W. (1982). Raman studies of pressure-induced frequency shifts in $d_0$- and $d_8$-napthalene, *J. Raman Spectrosc.* **13**, 247.
Buback, M., and Harfoush, A. A. (1983). Near infrared absorption of pure *n*-heptane between 5000–6500 cm$^{-1}$ to high pressures and temperatures, *Naturforsch A: Phys., Phys. Chem., Kosmophys.* **38**, 528.
McGuigan, S., Strange, J. H., and Chezeau, J. M. (1983). The temperature and pressure dependence of molecular motion in solid naphthalene studied by NMR, *Mol. Phys.* **49**, 275.
Sawamura, S., Schizawa, S. A., Suzuki, K., and Tamiguchi, Y. (1982). Effect of pressure on the hydrogen bond formation of benzoic acid in heptane, *23rd High Pressure Conf., Japan, November 18–20, 1982,* Abstr., p. 330.
Schmidy, C. G., Moore, D. S., Schiferl, D., and Shaner, J. W. (1983). Backward stimulated Raman scattering in shock compressed in benzene, *Phys. Rev. Lett.* **50**, 661.
Shimidzu, H., and Ohinishi, T. (1983). High-pressure Raman study of liquid molecular crystal at room temperature, $CS_2$, *Chem. Phys. Lett.* **99**, 507.
Eiling, A., Schilling, J. S., and Bach, H. (1981). High pressure studies of superconductivity of La-chalcogenides. *In* "Physics of Solids under High Pressures," pp. 385–396. North-Holland Publ., Amsterdam, Netherlands.
Fritsch, G., Willer, J., Wildermuth, A., and Luescher, E., (1982). Pressure dependence of the electrical resistivity of some metallic glasses, *J. Phys. F* **12**, 2965.
Gulino, D. A., Faulkner, L. R., and Drickamer, H. G. (1983). High pressure photoelectric studies of semiconductor electrolyte systems, *J. Appl. Phys.* **54**, 2483.
Howald, R. A., Moe, A. A., and Roy, B. N. (1983). The high pressure disproportionation of spinel, $Mg_2Al_2O_4$, *High. Temp. Sci.* **16**, 111.
Maugh, T. M., (1983). Number of organic superconductors grow, *Science* **222**, 606.
Mignot, J. M., and Wittig, J., (1981). Low temperature electrical resistivity of unstable valent $CeIn_3$, $CePd_3$ and YbCuAl up to 225 kbar. In "Physics of Solids under High Pressure," pp. 311–318. North-Holland Publ., Amsterdam, Netherlands.
Mitsubishi Elec. Corp. (1982). Pressure resistant planar semiconductor device, *Jpn. Tokyo Koho, JP,* **57**, 59.
Parkin, S. S., Coulon, C., Jerome, D., Fabre, J. M., and Giral, L. (1983). Substitution of TMTSeF with TMTTF in $(TMTSeF)_2ClO_4$: High pressure studies, *J. Phys. (Les Ulis, Fr.)* **44**, 603.

Parkin, S. S., Creuzet, F., Jerome, D., Fabre, J. M., and Beckgaard, K. (1983). Pressure temperature phase diagrams of several (TMTTF)$_2$X compounds, stabilization of a highly conducting metallic state under pressure in (TMTTF)$_2$Br, *J. Phys.* (*Les Ulis, Fr.*) **44,** 975.

Senapati, H., Parthasarathy, G., Lakshimikumar, S. T., and Rao, K. J. (1983). Effect of pressure on the fast ion conduction in silver iodide silver oxide molybdenum oxide, *Phil. Mag. B***47,** 291.

Williams, J. M., Beno, M. A., Sullivan, J. C., Banovetz, L. M., Braam, J. M., Blackman, G. S., Carlson, C. D., Greer, D. L., and Loesing, D. M. (1983). Design of organic metals based in tetramethyl-tetraselenafulvalene: Novel structural implications and predictions, *J. Am. Chem. Soc.* **105,** 643.

## Chapter 7. Special Applications

### Geochemical and Geophysical Applications

Hazen, R. M., and Finger, L. W. (1983). High pressure and high temperature crystallographic study of gillespite I–II phase transition, *Am. Mineral.* **68,** 595.

Ishidate, T. (1982). Pressure dependence of lattice constants of wurzite type crystal, *23rd High Presure Conf., Japan, November 18–20, 1982,* Abstr. p. 58.

Lacam, A. (1982). Study of electrical conductivity jump produced by olivine spinel transition in a laser heated diamond anvil cell up to 200 kbar, *C. R. Seances Acad. Sci. Sect. 2* **295,** 795.

Weng, K., Xu, J., Mao, H. K., and Bell, P. M. (1982–1983). Preliminary FT–IR spectral data on the SiO$_6^{8-}$ octahedral group in silicate-perovskites, *Annu. Rep. Geophys. Lab., Washington, D.C.* **82,** 355–359.

Xu, J., Mao, H. K., Weng, K., and Bell, P. M. (1982–1983). High-pressure, Fourier transform infrared spectra of forsterite and fayalite, *Annu. Rep. Geophys. Lab., Washington, D.C.* **82,** 350–352.

Xu, J., Mao, H. K., Weng, K., and Bell, P. M. (1982–1983). Preliminary data on FT–IR frequency shifts in hypersthene at high pressure, *Annu. Rep. Geophys. Lab., Washington, D.C.* **82** 352–353.

Yagi, T., Suzuki, T., and Akimoto, S. (1982). Static compression of FeO up to mbar range, *23rd High Pressure Conf., Japan, November 18–20, 1982,* Abstr. p. 50.

### Electrical Conduction Section

Alakhverdiey, K. R., Gasymov, S. G., and Salaev, E. Y. (1982). Effect of pressure on the electrical conductivity and Hall effect of thallium selenide, *Phys. Status Solidi A* **74,** K141.

Carlsson, A. E., and Ashcroft, N. W. (1983). Approaches for reducing the insulator metal transition pressure in hydrogen, *Phys. Rev. Lett.* **50,** 1305.

Cleaver, B., and Zani, P. (1983). The effect of pressure on the electrical conductivity of the molten halides of mercury and the molten iodides of cadmium, gallium and indium, *Z. Naturforsch A, Phys. Chem., Kosmophys.* **38,** 120.

Wudl, F., Nalewajek, D., Troup, J. M., and Extine, M. W. (1983). Electron density distribution in the organic superconductor (TMTSF)$_2$AsF$_6$, *Science* **222**, 415.

Zurawsky, W. P., Littman, J. E., and Drickamer, H. G. (1983). Pressure induced conduction and valence bond shifts in indium phosphide and gallium arsenide from measurements of semiconductor–electrolyte interfaces, *J. Appl. Phys.* **54**, 3216.

Zvarykina, A. V., Kushch, L. A., Laukhin, V. N., Lependina, O. L., Yanovskaya, I. M., and Yagubskii, E. B. (1983). Synthesis and electroconducting properties of TSeT$_2$Br$_x$Cl$_{1-x}$ mixed halides at normal and high pressure, *Izv. Akad. Nauk. SSSr. Ser. Khim.*, No. 7, 1625.

## Chapter 8. Miscellaneous Applications

Asaumi, K., Mori, T., and Kondo, Y. (1982). A research for metallic Xe by optical absorption, *23rd High Pressure Conf., Japan, November 18–20, 1982*, Abstr., p. 16.

Schiferel, D., Mills, R. L., and Trimmer, L. E. (1983). X-ray study of xenon to 23 GPa, *Solid State Commun.* **46**, 783.

Skelton, E. F., Qadri, S. B., Webb, A. W., Ingalls, R. G., and Traquadu, J. M. (1983). Pressure-induced disproportionation in cuprous bromide, *Phys. Lett. A* **94**, 441.

Smolander, K. J., On the high-pressure polymorphism of cuprous chloride, *J. Phys. C* **16**, 3673.

Wijngaarden, R. J., Goldman, V. V., and Silvera, I. F. (1983). Pressure dependence of the optical phonon in solid hydrogen and deuterium up to 230 kbar, *Phys. Rev. B* **27**, 5084.

# INDEX

## B

Band shapes with pressure, 4
Beam condensers, 17, 26, 27
Behavior classes for pressure-induced solid-state changes, 138–140
Brillouin scattering at high pressure, 29–32, 35, 36, 44

## C

Complementary techniques with DAC, 28–37
Cuprous chloride, 242–244
  possible decomposition with pressure, 243–244

## D

Diamond, 16, 17, 20, 26
Diamond anvil cell (DAC), 9–14, 40, 41
Disproportionation reactions with pressure
  cuprous chloride, 243–244

ferrous silicate spinel, 202
synthetic basalt glasses, 204–205
Doubling of vibrations with pressure, 5
Drickamer cell, 10

## E

EDXRD at high pressure, 36, 44
Electrical conductivity at high pressure, 28, 29, 32–34
Electrical conductivity of mantle, 197
Electrical conductor types
  Bechgaard salts, 218–219
  ion-polymers, 209
  Krogmann salts, 209
  linear 1-D complexes, 210–211
  metal phthalocyanines, 212–213
  organic donor–acceptor complexes, 208–209
  polyacetylene, 213–217
  polymeric, 209, 213
  $(SN)_x$, 216, 217
  TMTTF-TCNQ, 216–219
  $(TMTSF)_2ClO_4$, 218–219

Electrical conductor types
  (TMTSF)$_2$SbF$_6$, 218–219
  (TMTSF)$_2$TaF$_6$, 218–219
  transition metals, 208–209
Expansion of molecular volume, 5
Explosives at high pressure, 44, 184–187

**F**

Forensic science with DAC, 220–230
Frequency shifts with pressure, 3–4

**G**

Generation of pressure and temperature simultaneously, 28
Geochemical and geophysical applications of pressure, 194–208
Grüneisen parameters, 63–64, 92, 93

**H**

High-pressure transitions of Si minerals, Si IV → Si VI, 202–203
Hydrostatic pressures, 46–47
  use of frozen gases, 47
  use of liquid media, 46

**I**

Instrumentation used at high pressures, 9–44
Intensity changes with pressure, 4–5

**L**

Lattice modes
  longitudinal optical, 67–83
  transverse optical, 67–83
Laser
  He–Cd, 30
  He–Ne, 30
  Nd–YAG, 28–30
Lubricant studies at high pressure, 219–221

**M**

Megabar DAC, 9, 14, 15
Metallic hydrogen, 236–240
Metallic xenon, 241–242
Mixed crystals
  CdS$_{1-x}$, Se$_x$, 71–73
  K$_{1-x}$Rb$_x$I, 78, 80, 81
  ZnS$_{1-x}$Se$_x$, 70–74
Mössbauer spectroscopy at high pressure, 28, 34

**N**

Naturally occurring pressures, 7
Nickel dimethylglyoxime method of pressure calibration, 50–51
Nonhydrostatic pressure conditions, 46
Nonrigidity of solids at high pressure
  summary of interconversion of solids, 137–140

**O**

Optical instrumentation for high pressure, 17
Optical windows for high pressure, 16, 17–20

**P**

Piston–cylinder cell, 10, 11
Possible mineral distribution in mantle, 207–208
Pressure calibration, 45–62
  applied load method, 49, 50
  MgO:V$^{+2}$ fluorescence method, 58, 59
  nickel dimethylglyoxime method, 50, 51
  ruby scale, 51–58
  shifts of vibrations, 58, 59
  transducer method, 58
Pressure effects (general), 2–5
Pressure effects on coordination compounds, 120–160

effects on asymmetric and symmetric stretch, 155
ligand isomerism, 145, 146
oxidation–reduction, 144, 148
spin-state effects, 140–144
structural interconversion, 124–137
   five-coordinate, 130–133
   four-coordinate, 125–130
   seven- and eight-coordinate, 137
   six-coordinate, 133–137
   theory of solid-state conversion, 120–124
vibrational transitions, 146–155
Pressure effects on inorganic compounds
   $AB_2$ halides
      $CdX_2$, 89, 90
      $HgX_2$, 85–89
      $PbF_x$, 89, 90
   alkali metal cyanides, 94, 95
   alkali metal halides, 73–81
   alkaline earth halides, 81–83
   ammonium halides, 90–94
   azides ($NaN_3$, $KN_3$, $TlN_3$), 95–96
   bihalide salts, 84–85
   ionic conductors ($M_2HgI_4$), 107–113
   $K_2HPO_4$, 101
   nitrates and carbonates ($AgNO_3$, $CaCO_3$, $KNO_3$), 98–99
   nitrites,
      $KNO_2$, 97–98
      $NaNO_2$, 97
   organic ammonium salts, 114–115
   oxides (MO, $MO_2$)
      $Ag_2O$, 104
      $CO_2$, 104, 119
      $Cu_2O$, 104
      He (ice), 119
      $H_2O$ (ice), 105–107
      $MO_2$, 104
      $NH_3$ (ice), 119
      $SiO_2$, 102–104
      $TeO_2$, 101, 102
      $TiO_2$, 102
   $Rb_2HPO_4$, 101
   SbSI, 96–98
   silicon dioxide
      $\alpha$-cristabolite, 103, 105
      $\beta$-cristabolite, 103, 105
      fused silica, 105
      pyrex, 105
      $\alpha$-quartz, 103, 105
      $\beta$-quartz, 103, 105
      stishkovite, 103, 105
      tridymite, 103
      vycor, 105
   spinel,
      $Ba_2MoO_4$, 100
      $M'ReO_4$, 100
      $Ba_2WO_4$, 100
      $Na_2MoO_4$, 99
      $Na_2WO_4$, 99, 100
   thallium iodide, 83, 84
Pressure effects on organic and biological compounds, 163–191
   adamantanone and adamantane, 181, 182
   alkyl-substituted ammonium halides, 183, 184
   ammonium salts of organic acids, 182, 183
   $\alpha$-bacteriorhodopsin, 190, 191
   benzene, 169–172
   camphor, 185
   $CCl_4$, $CBr_4$, 165–167
   $p$-dihalobenzenes, 168–175
   dihalocyclohexanes, 185
   explosives, 44, 184–187
   $\alpha$-glycine, 191
   hydrogen-bonded compounds, 186–188
   iodoform, 164
   methane ice, 176–182
   methanol and ethanol, 165
   naphthalene and octafluoronaphthalene, 172–175, 177
   pyrazine, 184
   1,3,5-trioxane, 183
   1,3,5-trithiane, 183
Pressure effects on periclase and wüstite
   electrical conduction, 206, 207
   optical absorption, 206, 207
Pressure effects on physical properties
   charge-transfer reactions, 196
   coordination number, 197
   electrical conductivity, 197

magnetic properties, 197
oxidation state, 197
phase transition, 197
spin states, 197
structural response of minerals, 198–199
Pressure units, 5–6

**R**

Raman spectroscopy at high pressure, 21
Ruby pressure scale, 51–58

**S**

Sapphire, 17, 20, 26
Seismic velocity in mantle, 196
Shock-induced changes in mantle, 200–201

Spectrometers
  IR, 22, 23
  Raman, 24, 25

**T**

Types of mineral lattices, 199–200
  coordination, 199–200
  layer, 199–200
  metallic, 199–200

**V**

Viscosity at high pressure, 44

**X**

X-ray spectroscopy at high pressure, 28, 29, 31, 41–43